Sleep Disorders in Parkinson's Disease

Chun-Feng Liu

Editor

Sleep Disorders in Parkinson's Disease

Management and Case Study

Springer

Editor
Chun-Feng Liu
Department of Neurology
Second Affiliated Hospital of Soochow University
Suzhou, Jiangsu
China

ISBN 978-981-15-2480-6   ISBN 978-981-15-2481-3   (eBook)
https://doi.org/10.1007/978-981-15-2481-3

This Springer imprint is published by the registered company Springer Nature Singapore Pte Ltd.
The registered company address is: 152 Beach Road, #21-01/04 Gateway East, Singapore 189721, Singapore

Preface

Sleep disorders are common non-motor manifestations in Parkinson's disease (PD). In China, the prevalence of PD patients with sleep disturbance ranges from 47.66 to 89.10%. Sleep disorders usually have a significant negative impact on quality of life and may also serve as markers to identify patients in the preclinical stage of PD.

There are different types of sleep disturbances in patients with PD such as insomnia, excessive daytime sleepiness, rapid eye movement sleep behavior disorders, restless legs syndrome, and sleep-disordered breathing. Because PD has a variety of clinical manifestations, sleep disorders are most easy to be overlooked. Standardized assessment and management of PD patients with sleep disturbances will be beneficial to improve the overall quality of diagnosis, treatment, and prognosis.

The past decade has witnessed major advances in the understanding of sleep physiology and pharmacology. However, sleep problems of PD have their unique characteristics, either related to disease itself or drugs. In order to provide a better understanding of the mechanisms that underlie sleep disorders in PD, we are pleased to invite very experienced and senior experts to convert this evidence into *Management Recommendations on Sleep Disturbance of Patients with Parkinson's Disease*. I am grateful to them for their hard work and for their expertise. However, considering that these recommendations into practice might have been problematic, many case reports in this book would make learning and understanding more visual, lifelike, and convenient.

This book aims to provide a comprehensive overview to advances of PD-related sleep disorders and give a practical guidance to clinicians and scientists on management of sleep disorders in PD, not merely scientists and clinicians, even to present a complete knowledge structure to the general population.

We also found that there was a lack of large clinical randomized controlled trials. We hope that this book will not only aid current treatment of the disease, but will also stimulate efforts to improve future management more quickly than has been possible to date.

Suzhou, China   Chun-Feng Liu
December 2019

Acknowledgments

This book is supported by National Key R&D Program of China (2017YFC0909100) and Jiangsu Provincial social development projects (BE2018658).

Contents

27. Oishi Y, Xu Q, Wang L, Zhang BJ, Takahashi K, Takata Y, et al. Slow-wave sleep is controlled by a subset of nucleus accumbens core neurons in mice. Nat Commun. 2017;8:734. https://doi.org/10.1038/s41467-017-00781-4.

28. Yuan XS, Wang L, Dong H, Qu WM, Yang SR, Cherasse Y, et al. Striatal adenosine A2A receptor neurons control active-period sleep via parvalbumin neurons in external globus pallidus. elife. 2017;6:6. https://doi.org/10.7554/eLife.29055.

29. Saper CB, Fuller PM, Pedersen NP, Lu J, Scammell TE. Sleep state switching. Neuron. 2010;68(6):1023–42.

30. Saper CB, Scammell TE, Lu J. Hypothalamic regulation of sleep and circadian rhythms. Nature. 2005;437(7063):1257–63.

31. Sherin J, Shiromani P, McCarley R, Saper C. Activation of ventrolateral preoptic neurons during sleep. Science. 1996;271(5246):216–9.

32. Hara J, Beuckmann CT, Nambu T, Willie JT, Chemelli RM, Sinton CM, et al. Genetic ablation of orexin neurons in mice results in narcolepsy, hypophagia, and obesity. Neuron. 2001;30(2):345–54.

33. Chuah YL, Venkatraman V, Dinges DF, Chee MW. The neural basis of interindividual variability in inhibitory efficiency after sleep deprivation. J Neurosci. 2006;26(27):7156–62.

34. Muzur A, Pace-Schott EF, Hobson JA. The prefrontal cortex in sleep. Trends Cogn Sci. 2002;6(11):475–81.

35. Chee MW, Choo WC. Functional imaging of working memory after 24 hr of total sleep deprivation. J Neurosci. 2004;24(19):4560–7.

6. Koenigs M, Holliday J, Solomon J, Grafman J. Left dorsomedial frontal brain damage is associated with insomnia. J Neurosci. 2010;30(47):16041–3.

Hur EE, Zaborszky L. Vglut2 afferents to the medial prefrontal and primary somatosensory cortices: a combined retrograde tracing in situ hybridization. J Comp Neurol. 2005;483(3):351–73.

oshida K, McCormack S, España RA, Crocker A, Scammell TE. Afferents to the orexin urons of the rat brain. J Comp Neurol. 2006;494(5):845–61.

o H, Yokoi M. Striatal medium spiny neurons terminate in a distinct region in the lateral thalamic area and do not directly innervate orexin/hypocretin-or melanin-concentrating one-containing neurons. J Neurosci. 2007;27(26):6948–55.

I, Tanaka K, Chiba T. Efferent projections of the nucleus: biotinylated dextran amine study. Brain Res. 7(1):73–93.

hung S, Lu D-P, Cho YK. Descending projections from the nucleus accumbens ress activity of taste-responsive neurons in the hamster parabrachial nuclei. J ol. 2012;108(5):1288–98.

ahm D, Churchill L, Kalivas P, Wohltmann C. Specificity in the projection patterns ore and shell in the rat. Neuroscience. 1991;41(1):89–125.

Z, Luo YJ, Zhao YN, Sun HX, Yin D, et al. The rostromedial tegmental nucleus non-rapid eye movement sleep. PLoS Biol. 2018;16(4):e2002909. https://doi. nal.pbio.2002909.

H, Murphy JA, Burns J, Semba K. Differential c-Fos immunoreactivity in cell groups following systemic administration of caffeine in rats. J Comp):667–89.

Part I

General Concepts

An Overview of Parkinson's Disease

1

Cheng-Jie Mao and Chun-Feng Liu

Abstract

Parkinson's disease (PD) is the second most prevalent neurodegenerative disease in the nervous system after Alzheimer's disease (AD).

Although the importance of non-motor symptoms is now widely acknowledged, we are still hampered by a lack of well-conducted research into effective treatments.

Regardless of certain possible pathogenesis, including aging, environmental toxicant, and heredity, the authentic cause of PD has still been remained uncovered so far. Pre-warning screening of PD before motor symptoms begin is one of the scientific hot issues and also an essential way to early diagnosis in need for thorough exploration.

In addition, progression of PD may depend on many unidentified pathogenic factor. Screening for early warnings may be discovered and enabled by other sensitive markers. There is a long way to go for discovering and establishing other sensitive and specific markers for early screening. It remains uncovered how we can know about physiological states and how we can create a better individualized way of screening depending on various group of people at risk applied to very early stage.

Keywords

Parkinson's disease · Pathogenesis · Premotor stage · Non-motor symptoms · Pre-warning screening

C.-J. Mao · C.-F. Liu (✉)
Department of Neurology, The Second Affiliated Hospital of Soochow University,
Suzhou, China
e-mail: liuchunfeng@suda.edu.cn

Parkinson's disease (PD) is the second most prevalent neurodegenerative disease in the nervous system after Alzheimer's disease (AD), which is caused by α-synuclein aggregation, leading to subsequent dopaminergic degeneration in substantia nigra and striatum. More than 75% of dopamine neurons have died when motor symptoms occurred. The first detailed description of 6 PD patients by James Parkinson centuries ago consists of both motor symptoms, with tremor, rigidity, bradykinesia, postural instability included, and non-motor symptoms, which contains fatigue, somnipathy, hypersalivation, urinary and defecating disorder, etc.

Regardless of certain possible pathogenesis, including aging, environmental toxicant, and heredity, the authentic cause of PD has still been remained uncovered so far. In fact, according to Braak's staging of sporadic Parkinson's disease, changes have begun to be witnessed on dopaminergic neurons almost 10 years before motor symptoms rises.

On stage 1, patients begin to show hyposmia with their olfactory bulb and anterior olfactory nucleus degenerated. On stage 2, pathogenic change can be diagnosed in dorsal raphe nucleus, nuclei pontis, locus coeruleus, and other nucleus in lower brainstem with autonomic and sleep disorders manifested. Patients tend to suffer from typical symptoms like tremor and rigidity on stage 3 and 4 for the losses of dopaminergic neurons in mesencephalon. Lastly, patients on stage 5 and 6, when Lewy body deposits in limbic system and cortex, are likely to have depression, impaired cognition, and neuropsychiatric symptoms.

It is usually referred to as the prodromal stage before the stage 3, that is, the stage 1 and 2, before the appearance of typical motors symptoms. At prodromal phase, during which patients suffer from hyposmia, sleep disorders, depression, constipation, and other non-motor symptoms. We normally take those people as PD risk category. At early motor stage, after which four main typical symptoms commence, the dominant disorders contain fatigue, pain, diplopia, etc. Some of these non-motor symptoms are caused by dopaminergic denervation, while some are not. For instance, hyposmia and cognitive dysfunction are related to degeneration of cholinergic system, and sleep disorders are linked to hypocretin and serotonin.

Three primary motor symptoms of PD are tremor, rigidity, and bradykinesia (slowness of movement). Some patients have difficulty with walking and gait, but not in the early stage of PD. It should be noticed that not all of these symptoms must exist for the diagnosis of PD. In fact, especially in the early stages of PD, many people may only notice one or two of these motor symptoms. For the diagnosis of PD, bradykinesia is necessary. Not everyone with PD has tremor, nor is tremor a proof of PD. So as rigidity. The motor symptoms characteristically start in one hand, foot, or leg and ultimately affect both sides of the body.

Bradykinesia is a frequent symptom in PD. In addition to the slowness of movement, the mask-like face expression and the decreased blink rate of eyes are often found in PD patients. The tremor of PD often occurs at rest with a rhythm of 4–6 Hz. Except for limbs, tremor can also occur in the jaw, mouth, or tongue. Rigidity refers to a tightness of the limbs or neck. In the early stages of PD, rigidity may be wrongly thought to be caused by arthritis or lumbar disc protrusion. Some patients even take the surgical treatment of lumbar disc herniation. Postural instability is more common in the later stages of PD. The inability to maintain steady, upright posture made them easy to fall down.

People with Parkinson's disease often, for example, experienced many changes in their mood, behavior, cognition, sleep, sense of smell, and gastrointestinal and blood pressure control, and otherwise unexplained pain. These non-motor symptoms often appear many years before the onset of motor symptoms. Two examples of this are the impairment in olfaction and a sleep disorder called REM sleep behavior disorder, which may occur more than a decade before any discernible physical change. Numerous studies have shown that non-motor symptoms are often very important for the quality of life of patients with Parkinson's disease and their families.

Because PD is a type of movement disorder, the non-motor symptoms are often been overlooked by the patients and doctors. The non-motor symptoms can be divided into several groups: disturbances of sense (smell and pain), sleep problems (insomnia, REM sleep behavior disorder), motion disorders (depression and anxiety), psychosis, fatigue, cognitive changes, urinary issues, gastrointestinal issues (constipation), sweating, sexual concerns, and eye and vision issues.

More and more people pay attention to the non-motor symptoms, which are not only the common clinical symptoms but also the biomarker for Parkinson's disease. Among the sleep problems that are commonly experienced by PD patients are the inability to fall asleep and vivid dreams.

In 2015, Movement Disorder Society attached great importance to non-motor symptoms when enacting diagnostic criteria of Parkinson's disease, who proposed hyposmia and cardiac sympathetic denervation as supportive criteria and orthostatic hypotension in the first 5 years, severe urinary retention, and urinary incontinence as red flags. Similarly, the absence of somnipathy, autonomic disorders, or psychological disorder as depression in the first 5 years is also regarded as a red flag.

Although the importance of non-motor symptoms is now widely acknowledged, we are still constricted by a lack of well-conducted research into effective treatments.

PD has diffident genetic and biochemical biomarkers. Biomarkers are objective measures that serve as indicators of normal biological processes, pathogenic processes, or pharmacologic responses to therapeutic interventions. It has promise to delineate molecularly defined subgroups of PD patients who may be most likely to benefit from specific therapeutic interventions. A single biomarker cannot reflect the complexity of PD. Clinical laboratory, imaging, and genetic factors should be combined to predict the onset and progression of PD accurately. Genetic mutations such as α-syn (*SNCA*), *Parkin*, *PTEN-induced kinase 1* (*PINK1*), *DJ-1*, and *leucine-rich repeat kinase 2* (*LRRK2*) account for 2–3% of all cases with classical parkinsonism, which is often clinically indistinguishable from idiopathic PD. Twenty other risk loci were identified by population-based genome-wide association studies (GWAS). The genetic cause of PD may be present in individuals several years before clinical symptoms and often for decades. The presence of clinical symptoms, including both motor and non-motor characteristics, is the most important diagnostic marker for PD. Several non-motor symptoms (RBD, such as olfactory dysfunction, depression, and bowel dysfunction) usually precede the motor features of PD. RBD is particularly associated with subsequent development of Parkinson's disease and dementia. Hyposmia and depression is present about 90% PD patients. Considering the biochemical markers, α-syn can be detected in CSF, saliva, serum, urine, and also the gastrointestinal tract. Inconsistent findings suggest that the level of α-syn in the cerebrospinal fluid of PD patients may be lower than in healthy controls. And several neuroimaging techniques have been developed and can be used to support the clinical diagnosis of PD. There is still much work to be done to find the right PD biomarker. In the last decade, new powerful methods have emerged in proteomics, metabolomics, and transcriptomics to identify small changes in protein, metabolites, or RNA profiles in tissue or fluids from healthy and diseased individuals. Priority should be given to studies that assessment of combinations of clinical, genetics, biochemical markers, and imaging.

At present, there is only a small proportion of PD patients with known mutations of pathogenic genes and loci, and the genetic background of most of the patients is not clear. Most of the genetic factors of PD patients are not clear at present. The autosomal dominant genes of PD patients include *SNCA, LRRK2, VPS35, UCH-L1, HTRA2, GIGYF2, EIF4G1, CHCHD2, TMEM230, DNAJC13*, and *LRP10*. According to race, family history and age at onset, about 0.1–30% of PD patients suffered from autosomal dominant mutation. The identified autosomal recessive pathogenic genes of PD include *Parkin, PINK1, DJ-1 (PARK7), ATP13A2, PLA2G6, FBXO7, DNA JC6, SYNJ 1*, and *VPS13C*. These autosomal recessive genes account for about 13% of patients with early onset PD (onset age less than 40 years old). *RAB39B* is the pathogenic X-linked gene of PD. With the development and maturity of next generation sequencing (NGS), more PD-related genes and mutations will be found in the future.

Pre-warning screening of PD before motor symptoms begin is one of the scientific issues and also an essential way to early diagnosis in need for thorough exploration. It is extremely important for early diagnosis to realize the relationship between the time when non-motor symptoms appear and dopaminergic dysfunction in striatum. Hyposmia, constipation, and sleep disorder, especially rapid eye movement behavior disorders (RBD), can appear at the time when dopaminergic degeneration begins.

The relationship between idiopathic RBD (iRBD) and neurological diseases is well established, and iRBD has attracted increasing attention as patients may eventually be diagnosed with parkinsonism, such as PD, multiple-system atrophy (MSA), or dementia with Lewy bodies (DLB). IRBD has been used as a premotor symptom marker for neurodegenerative disorders. It may precede PD motor manifestations or develop after PD onset. Some studies have explored the conversion risk of Parkinsonism from iRBD. A longitudinal, prospective study of iRBD has revealed that the 5-year conversion risk of neurodegenerative disease is 17.7%, the 10-year risk 40.6%, and the 12-year risk 52.4%. The probability of conversion increases with the combination of other non-motor symptoms, especially olfactory dysfunction and constipation. The sleep disorders and depression may occur before the onset of motor symptoms, and impairment of cognition appears at last. The conversion risk is also associated with the gene analyses and imaging results. More attention should be paid to study the biomarkers of PD.

In the most circumstances, individuals of high PD risk have to take screening tests over and over again for early signs depending on the progression rate of α-synucleinopathy. Frequent screening is necessary if rapidly progressive, and vice versa.

In addition, progression of PD may depend on many unidentified pathogenic factor. Screening for early warnings may be discovered and enabled by other sensitive markers. There is a long way to go for discovering and establishing other sensitive and specific markers for early screening. It remains uncovered how we can know about physiological states and how we can create a better individualized way of screening depending on various group of people at risk applied to very early stage.

An Overview of Roles of the Basal Ganglia in Sleep-Wake Regulation

2

Wei-Min Qu, Ze Zhang, Huan-Ying Shi, and Zhi-Li Huang

Abstract

Sleep disorders are frequent in Parkinson's disease and their prevalence increases with disease progression. Previous studies demonstrated that sleep disorders could appear as an initial manifestation of Parkinson's disease even decades before motor signs, which highlight their clinical association in these early stages. Dysfunction of dopaminergic transmission in the basal ganglia involves in the pathogenesis of Parkinson's disease and sleep disorders. This chapter will focus on reviewing the role of the basal ganglia in control of sleep and wakefulness.

Keywords

Basal ganglia · Dopamine · Adenosine · Sleep

2.1 Introduction

Sleep disorders are frequent in Parkinson's disease (PD) and could be an initial manifestation of PD even decades before motor signs. The symptoms of sleep disorders in PD patients include night insomnia, increased sleepiness, sleep fragmentation, reduced sleep efficiency, and rapid eye movement (REM) sleep behavior disorder, thus having a serious impact on the patient's sleep quality and

W.-M. Qu · Z. Zhang · H.-Y. Shi · Z.-L. Huang (✉)
Department of Pharmacology, School of Basic Medical Sciences,
Fudan University, Shanghai, China

MOE Frontiers Center for Brain Science, Fudan University, Shanghai, China

Institutes of Brain Science and State Key Laboratory of Medical Neurobiology,
Fudan University, Shanghai, China
e-mail: huangzl@fudan.edu.cn

© Springer Nature Singapore Pte Ltd. 2020
C.-F. Liu (ed.), *Sleep Disorders in Parkinson's Disease*,
https://doi.org/10.1007/978-981-15-2481-3_2

increasing the risk of dementia. In recent years, new hypotheses involving dopaminergic and non-dopaminergic mechanisms emerged concerning sleep-wake neurobiology in PD. Over the past decades, notable advances have been made in the understanding of the role of the dopaminergic system in circadian mechanisms and sleep-wake physiology [1]. The loss of dopaminergic transmission in PD severely affects the function of the basal ganglia (BG), which play a critical role in sleep-wake regulation [2]. As a consequence, the dysfunction of the BG can cause sleep disorders in PD.

Dopamine D_2 receptors (D_2Rs) are highly expressed on striatopallidal neurons in the indirect pathway of the BG, in which adenosine A_{2A} receptors (A_{2A}Rs) are co-expressed with D_2Rs. Caffeine, an A_1 receptors (A_1Rs) and A_{2A}Rs antagonist, has been indicated to induce wakefulness via blockade of adenosine function on A_{2A}Rs [3]. The arousal effect of caffeine could also be modulated by A_{2A}Rs in the shell of the nucleus accumbens (NAc) [4]. In this chapter, we attempt to review the existing evidence regarding adenosine and dopamine in the BG for regulating sleep-wake cycle and discuss anatomical and molecular models in understanding mechanisms of sleep-wake regulation.

2.2   The Roles of the BG in the Sleep-Wake Cycle

The BG consist of the striatum, globus pallidus (GP), subthalamic nucleus (STN), and substantia nigra (SN) [5]. The BG has a strong connection with dopaminergic neurons in the midbrain and cohesively involved in optimizing behavior and regulating the vigilance state of wakefulness. Results from neurotoxic lesion of the striatum indicate an obvious causal association between striatal structures and sleep-wake cycle modulation, that is, the dorsal striatum contributes to wakefulness and the NAc benefits sleep [6, 7]. These studies revealed that bilateral lesions in the striatum can cause reduction and fragmentation wakefulness time, in which effect could be attenuated when NAc was additionally included. In contrast, selective ablation of NAc resulted in an increase of wakefulness and reduction of duration on bouts of non-rapid eye movement (non-REM, NREM) sleep.

Lesions to external GP (GPe), surprisingly, contributed to increase in wakefulness and pronounced fragmentation of NREM sleep and wakefulness [6] while lesions of invading internal GP or STN failed to exhibit altered sleep-wake [6]. Loss of DA-ergic neurons and disturbed REM sleep (REMS) are associated with PD. Recent study also demonstrated that SN-DA-ergic neurons act on the SN-GABA-ergic to regulate REMS and loss of neurons in the SN can lead to increased wakefulness [8]. Interestingly, a generalized slowing of the cortical activity, either during wakefulness or sleep, was observed in rat models with lesions in the CPu, NAc, and GPe, recorded by electroencephalogram. It is hypothesized dorsal striatum regulates sleep-wake cycle and cortical activation via the potential dorsostriato-pallido-cortical loop [6, 9]. More precisely, GABAergic neurons in the caudate-putamen (CPu) projects to the GPe, which in turn projects directly to the cerebral cortex. GABAergic neurons in the adjacent basal forebrain (BF) project to cortical inhibitory interneurons and thus promote wakefulness, whereas those in the interior of the GPe inhibit pyramidal cells and promote sleep [6].

2.3   The Role of the Dopaminergic and Adenosinergic System in the BG in the Sleep-Wake Cycle

Although mesolimbic dopamine system from the midbrain to the striatum has been extensively focused [10, 11], experimental studies shed important light on the roles of adenosine and dopamine in the BG in regulating sleep-wake cycle. *In vivo* microdialysis experiments combined polysomnographic recording revealed that levels of extracellular dopamine was increased during wakefulness and REM sleep in the medial prefrontal cortex (mPFC) and NAc [12]. Fiber photometry coupled with polysomnographic recordings demonstrated that dorsal striatum dopamine levels correlated with the spontaneous sleep-wake cycle, and striatal dopamine levels were at their highest during wakefulness, lower during NREM sleep, and lowest during REM sleep and are correlated with sleep-state transitions [13]. Animal study showed that the deletion of D_2Rs led to significant decrease in wakefulness, along with an increase in NREM and REM sleep [14], indicating that the crucial role of D_2R in maintaining wakefulness. When administration of agents targeting NAc, application of D_2R agonist increased wakefulness while D_2R antagonist promoted sleep [15]. Studies using positron emission tomography also showed downregulation of D_2Rs expression in humans with sleep deprivation [16]. However, use of D_2R agonists, such as piribedil and pramipexole, clinically helping to relieve symptoms of PD and related disease, can sometimes cause sudden sleep attacks or sleepiness [17, 18]. Therefore, the function of D_2R agonists could be complex as they not only activate D_2Rs on striatal neurons but also reduce dopamine release in mesolimbic and mesocortical systems [19, 20].

Modafinil, a medication to promote wakefulness, can increase extracellular concentration of dopamine in the NAc and mPFC [21]. However, such effect was not observed in dopamine transporter knockout mice [22]. Further study using D_2R knockout mice plus dopamine D_1 receptor (D_1R) antagonist indicates that the arousal effect of modafinil could primarily mediated by D_2R [23]. A recent finding by using fiber photometry and simultaneously collecting polysomnographic recordings in freely behaving mice after environmental or pharmacological manipulations found that the wake-promoting agent modafinil, but not caffeine, induced the release of striatal dopamine [13].

Conversely, the administration of CGS21680, a highly selective $A_{2A}R$ agonist, to the subarachnoid space under the rostral BF produces c-fos expression within the shell of the NAc, the medial portion of the olfactory tubercle, and the ventrolateral preoptic area (VLPO) [24]. It is possible that the activation of c-fos in the VLPO was secondary to the effect of activation of $A_{2A}Rs$ in the NAc. Moreover, the direct infusion of the same $A_{2A}R$ agonist into the shell portion of the NAc induces NREM and REM sleep [25]. These observations indicate that $A_{2A}Rs$ in or close to the shell portion of the NAc promote sleep.

Caffeine is the world's most recognized agent to enhance wakefulness. Studies using global genetic knockout models found that $A_{2A}R$, not A_1R, mediates its arousal effect [3]. Using powerful tools for site-specific gene manipulations [4], we previously reported that selective deletion of the $A_{2A}Rs$ in the NAc shell can abrogate the effect caffeine-induced wakefulness. The antagonist effect of caffeine is premised on the tonically activated excitatory $A_{2A}Rs$ by adenosine. Given the availability of adenosine under the most basal conditions and the high expression of $A_{2A}Rs$ in the striatum, such tonic activation is feasible in the NAc shell. Thus, activation of the GABAergic output neurons by $A_{2A}Rs$ ultimately results in decreased activity of arousal systems in the cerebral cortex. In fact, after designer receptors are exclusively activated by a designer drug (DREADD) into the NAc of transgenic mice, in which Cre-recombinase is expressed under the $A_{2A}R$ promoter [26], NREM sleep can be induced during selective activation of NAc A_{2A} neurons [27]. In contrast, NAc dopamine D_1R-expressing neurons are essential for behavioral arousal [9]. Optogenetic activation of NAc D_1R neurons induces immediate transitions from NREM sleep to wakefulness, and chemogenetic stimulation prolongs arousal. Midbrain and lateral hypothalamus serve as the functional downstream [9]. Moreover, we recently demonstrate that striatal adenosine $A_{2A}R$ neurons control active-period sleep via parvalbumin (PV) neurons in external globus pallidus (GPe). We showed that chemogenetic activation of $A_{2A}R$ neurons in specific subregions of the striatum induced a remarkable increase in NREM sleep [28]. Anatomical mapping and immunoelectron microscopy revealed that striatal $A_{2A}R$ neurons innervated the GPe in a topographically organized manner and preferentially formed inhibitory synapses with GPe PV neurons. Moreover, lesions of GPe PV neurons abolished the sleep-promoting effect of striatal $A_{2A}R$ neurons. In addition, chemogenetic inhibition of striatal $A_{2A}R$ neurons led to a significant decrease of NREM sleep at active period, but not inactive period of mice.

2.4 A Model of NAc Involvement in Sleep-Wake Regulation

Several models have been postulated to interpret the regulation of sleep-wake network. Apart from inhibition of acetylcholine release in the BF in regulating sleep-wake cycle [29], a contemporary model describes a "flip-flop" arrangement. In this proposed model, sleep is induced by sleep-promoting neurons activation in the VLPO and wake-promoting neurons reciprocal suppression in the brainstem and hypothalamus [30–32]. During wakefulness, histaminergic neurons in the tuberomammillary nucleus, noradrenergic neurons in the locus coeruleus, and serotonergic neurons in the dorsal raphe nucleus have excitatory effect on arousal systems and thus inhibit the sleep-promoting neurons of the VLPO, which further inhibits of the arousal-promoting regions through GABAergic and galaninergic projections. Additionally, the "flip-flop" switch model supposes that orexin neurons of the lateral hypothalamus (LHA) help maintain wakefulness by suppressing unwanted transitions into sleep [33].

Given the increasing understanding of dopamine, adenosine, and glutamate, various functions of NAc in integrating locomotion and motivation-related behavior (e.g., dopaminergic inputs, emotional information from the amygdala, and executive/cognitive content from the prefrontal cortex) could be dissociable and explainable at neurotransmitter or neuromodulator levels. To date, emerging evidence suggests that the NAc is capable of regulating sleep and wakefulness through inhibition of neuronal populations in the ventral pallidum, the LHA, the parabrachial nucleus (PB), and the ventral tegmental area (VTA). Thalamus and mPFC, the circuit originating from the ventral pallidum, are uniquely sensitive to sleep and sleep need [34–37]. The orexinergic and glutamatergic neurons in the LHA not only send major projections to the BF and cerebral cortex but are also reciprocally connected to the NREM/wake flip-flop switch [30, 38–40]. The NAc shell sends projections to the PB [41, 42], which is an important component of the ascending arousal system. The NAc projects to the medial part of the VTA with a field of cortically projecting glutamatergic neurons [38], which are likely the tail end of a larger group of neurons of the supramammillary nucleus. The rostromedial tegmental nucleus (RMTg), also known as GABAergic tail of the VTA, exerts major inhibitory control over the DAergic system to promote NREM sleep [43]. To be noted, caffeine is believed to induce c-fos expression in non-dopaminergic neurons of the medial VTA [44], but whether the VTA/supramam-millary nucleus cell group relays the waking stimulus from the NAc to the cerebral cortex is unknown.

In conclusion, the above results provide several lines of evidence to understand physiological sleep-wake cycle, to explore the mechanisms of sleep disturbances in PD, and help to develop effective drugs for treating insomnia patients. Future studies are needed to identify the exact output projections of the striatum and the NAc that relays the waking stimulus from the BG to the sleep-wake regulatory network and eventually, leading to cortical awakening.

References

1. Videnovic A, Golombek D. Circadian and sleep disorders in Parkinson's disease. Exp Neurol. 2013;243:45–56. https://doi.org/10.1016/j.expneurol.2012.08.018.
2. Lazarus M, Huang ZL, Lu J, Urade Y, Chen JF. How do the basal ganglia regulate sleep-wake behavior? Trends Neurosci. 2012;35(12):723–32. https://doi.org/10.1016/j.tins.2012.07.001.
3. Huang Z-L, Qu W-M, Eguchi N, Chen J-F, Schwarzschild MA, Fredholm BB, et al. Adenosine A2A, but not A1, receptors mediate the arousal effect of caffeine. Nat Neurosci. 2005;8(7):858–9.
4. Lazarus M, Shen HY, Cherasse Y, Qu WM, Huang ZL, Bass CE, et al. Arousal effect of caffeine depends on adenosine A2A receptors in the shell of the nucleus accumbens. J Neurosci. 2011;31(27):10067–75. https://doi.org/10.1523/JNEUROSCI.6730-10.2011.
5. Crittenden JR, Graybiel AM. Basal ganglia disorders associated with imbalances in the striatal striosome and matrix compartments. Front Neuroanat. 2011;5:59.
6. Qiu MH, Vetrivelan R, Fuller PM, Lu J. Basal ganglia control of sleep–wake behavior and cortical activation. Eur J Neurosci. 2010;31(3):499–507.

7. Qiu M-H, Liu W, Qu W-M, Urade Y, Lu J, Huang Z-L. The role of nucleus accumbens core/ shell in sleep-wake regulation and their involvement in Modafinil-induced arousal. PLoS One. 2012;7(9):e45471.
8. Yadav RK, Khanday MA, Mallick BN. Interplay of dopamine and GABA in substantia nigra for the regulation of rapid eye movement sleep in rats. Behav Brain Res. 2019;376:112169. https://doi.org/10.1016/j.bbr.2019.112169.
9. Luo YJ, Li YD, Wang L, Yang SR, Yuan XS, Wang J, et al. Nucleus accumbens controls wakefulness by a subpopulation of neurons expressing dopamine D1 receptors. Nat Commun. 2018;9(1):1576. https://doi.org/10.1038/s41467-018-03889-3.
10. Ikemoto S. Dopamine reward circuitry: two projection systems from the ventral midbrain to the nucleus accumbens–olfactory tubercle complex. Brain Res Rev. 2007;56(1):27–78.
11. Sesack SR, Grace AA. Cortico-basal ganglia reward network: microcircuitry. Neuropsychopharmacology. 2009;35(1):27–47.
12. Lena I, Parrot S, Deschaux O, Muffat-Joly S, Sauvinet V, Renaud B, et al. Variations in extracellular levels of dopamine, noradrenaline, glutamate, and aspartate across the sleep–wake cycle in the medial prefrontal cortex and nucleus accumbens of freely moving rats. J Neurosci Res. 2005;81(6):891–9.
13. Dong H, Wang J, Yang YF, Shen Y, Qu WM, Huang ZL. Dorsal striatum dopamine levels fluctuate across the sleep-wake cycle and respond to salient stimuli in mice. Front Neurosci. 2019;13:242. https://doi.org/10.3389/fnins.2019.00242.
14. Qu W-M, Xu X-H, Yan M-M, Wang Y-Q, Urade Y, Huang Z-L. Essential role of dopamine D2 receptor in the maintenance of wakefulness, but not in homeostatic regulation of sleep, in mice. J Neurosci. 2010;30(12):4382–9.
15. Barik S, de Beaurepaire R. Dopamine D3 modulation of locomotor activity and sleep in the nucleus accumbens and in lobules 9 and 10 of the cerebellum in the rat. Prog Neuro-Psychopharmacol Biol Psychiatry. 2005;29(5):718–26.
16. Volkow ND, Tomasi D, Wang G-J, Telang F, Fowler JS, Logan J, et al. Evidence that sleep deprivation downregulates dopamine D2R in ventral striatum in the human brain. J Neurosci. 2012;32(19):6711–7.
17. Tan E. Piribedil-induced sleep attacks in Parkinson's disease. Fundam Clin Pharmacol. 2003;17(1):117–9.
18. Lipford MC, Silber MH. Long-term use of pramipexole in the management of restless legs syndrome. Sleep Med. 2012;13(10):1280–5.
19. Monti JM, Hawkins M, Jantos H, D'Angelo L, Fernández M. Biphasic effects of dopamine D-2 receptor agonists on sleep and wakefulness in the rat. Psychopharmacology. 1988;95(3):395–400.
20. Sebban C, Zhang X, Tesolin-Decros B, Millan M, Spedding M. Changes in EEG spectral power in the prefrontal cortex of conscious rats elicited by drugs interacting with dopaminergic and noradrenergic transmission. Br J Pharmacol. 1999;128(5):1045–54.
21. Murillo-Rodríguez E, Haro R, Palomero-Rivero M, Millán-Aldaco D, Drucker-Colín R. Modafinil enhances extracellular levels of dopamine in the nucleus accumbens and increases wakefulness in rats. Behav Brain Res. 2007;176(2):353–7.
22. Wisor JP, Nishino S, Sora I, Uhl GH, Mignot E, Edgar DM. Dopaminergic role in stimulant-induced wakefulness. J Neurosci. 2001;21(5):1787–94.
23. Qu W-M, Huang Z-L, Xu X-H, Matsumoto N, Urade Y. Dopaminergic D1 and D2 receptors are essential for the arousal effect of modafinil. J Neurosci. 2008;28(34):8462–9.
24. Scammell T, Gerashchenko D, Mochizuki T, McCarthy M, Estabrooke I, Sears C, et al. An adenosine A2a agonist increases sleep and induces Fos in ventrolateral preoptic neurons. Neuroscience. 2001;107(4):653–63.
25. Satoh S, Matsumura H, Koike N, Tokunaga Y, Maeda T, Hayaishi O. Region-dependent difference in the sleep-promoting potency of an adenosine A2A receptor agonist. Eur J Neurosci. 1999;11(5):1587–97.
26. Durieux PF, Bearzatto B, Guiducci S, Buch T, Waisman A, Zoli M, et al. D2R striatopallidal neurons inhibit both locomotor and drug reward processes. Nat Neurosci. 2009;12(4):393–5.

27. Oishi Y, Xu Q, Wang L, Zhang BJ, Takahashi K, Takata Y, et al. Slow-wave sleep is controlled by a subset of nucleus accumbens core neurons in mice. Nat Commun. 2017;8:734. https://doi.org/10.1038/s41467-017-00781-4.
28. Yuan XS, Wang L, Dong H, Qu WM, Yang SR, Cherasse Y, et al. Striatal adenosine A2A receptor neurons control active-period sleep via parvalbumin neurons in external globus pallidus. elife. 2017;6:6. https://doi.org/10.7554/eLife.29055.
29. Saper CB, Fuller PM, Pedersen NP, Lu J, Scammell TE. Sleep state switching. Neuron. 2010;68(6):1023–42.
30. Saper CB, Scammell TE, Lu J. Hypothalamic regulation of sleep and circadian rhythms. Nature. 2005;437(7063):1257–63.
31. Sherin J, Shiromani P, McCarley R, Saper C. Activation of ventrolateral preoptic neurons during sleep. Science. 1996;271(5246):216–9.
32. Hara J, Beuckmann CT, Nambu T, Willie JT, Chemelli RM, Sinton CM, et al. Genetic ablation of orexin neurons in mice results in narcolepsy, hypophagia, and obesity. Neuron. 2001;30(2):345–54.
33. Chuah YL, Venkatraman V, Dinges DF, Chee MW. The neural basis of interindividual variability in inhibitory efficiency after sleep deprivation. J Neurosci. 2006;26(27):7156–62.
34. Muzur A, Pace-Schott EF, Hobson JA. The prefrontal cortex in sleep. Trends Cogn Sci. 2002;6(11):475–81.
35. Chee MW, Choo WC. Functional imaging of working memory after 24 hr of total sleep deprivation. J Neurosci. 2004;24(19):4560–7.
36. Koenigs M, Holliday J, Solomon J, Grafman J. Left dorsomedial frontal brain damage is associated with insomnia. J Neurosci. 2010;30(47):16041–3.
37. Hur EE, Zaborszky L. Vglut2 afferents to the medial prefrontal and primary somatosensory cortices: a combined retrograde tracing in situ hybridization. J Comp Neurol. 2005;483(3):351–73.
38. Yoshida K, McCormack S, España RA, Crocker A, Scammell TE. Afferents to the orexin neurons of the rat brain. J Comp Neurol. 2006;494(5):845–61.
39. Sano H, Yokoi M. Striatal medium spiny neurons terminate in a distinct region in the lateral hypothalamic area and do not directly innervate orexin/hypocretin-or melanin-concentrating hormone-containing neurons. J Neurosci. 2007;27(26):6948–55.
40. Usuda I, Tanaka K, Chiba T. Efferent projections of the nucleus accumbens in the rat with special reference to subdivision of the nucleus: biotinylated dextran amine study. Brain Res. 1998;797(1):73–93.
41. Li C-S, Chung S, Lu D-P, Cho YK. Descending projections from the nucleus accumbens shell suppress activity of taste-responsive neurons in the hamster parabrachial nuclei. J Neurophysiol. 2012;108(5):1288–98.
42. Heimer L, Zahm D, Churchill L, Kalivas P, Wohltmann C. Specificity in the projection patterns of accumbal core and shell in the rat. Neuroscience. 1991;41(1):89–125.
43. Yang SR, Hu ZZ, Luo YJ, Zhao YN, Sun HX, Yin D, et al. The rostromedial tegmental nucleus is essential for non-rapid eye movement sleep. PLoS Biol. 2018;16(4):e2002909. https://doi.org/10.1371/journal.pbio.2002909.
44. Deurveilher S, Lo H, Murphy JA, Burns J, Semba K. Differential c-Fos immunoreactivity in arousal-promoting cell groups following systemic administration of caffeine in rats. J Comp Neurol. 2006;498(5):667–89.

Part II

Classification of Sleep Disorders in Parkinson's Disease

REM Sleep Behavior Disorder (RBD)

3

Tao Wang

Abstract

REM sleep behavior disorder (RBD) is a parasomnia detected by vPSG showing loss of muscle atonia during REM sleep. Patients with RBD could easily recollect dream occasions associated with movements. Over the past decade, researchers brought up the theory that RBD is a prodromal stage of Parkinson's disease (PD) and often indicates a worse prognosis among PD and RBD comorbid patients.

The diagnosis of definite RBD is specified by International Classification of Sleep Disorders-3 (ICSD-3), where PSG plays an essential role. RBD sometimes could be misdiagnosed as other sleep disorders such as NREM parasomnia, nocturnal panic attacks, nocturnal seizures, and nocturnal wandering associated with dementia, where PSG is critical in the differentiation.

Treatment of RBD is comprised of two parts, namely, injury risk reductions and medications. Although clonazepam and melatonin are commonly used, it should be noted that clonazepam has side effects on those with OSA or the elder with dementia.

Keywords

RBD · REM sleep · vPSG · Parkinson's disease (PD) · Parasomnia · Diagnosis · Treatment · Differentiation

T. Wang (✉)
Department of Neurology, Union Hospital, Tongji Medical College, Huazhong University of Science and Technology, Wuhan, People's Republic of China
e-mail: wangtaowh@hust.edu.cn

© Springer Nature Singapore Pte Ltd. 2020
C.-F. Liu (ed.), *Sleep Disorders in Parkinson's Disease*,
https://doi.org/10.1007/978-981-15-2481-3_3

REM sleep behavior disorder (RBD) is characterized by the loss of muscle atonia during REM sleep that could be associated with nightmares or active behaviors during dreaming, leading to disturbed sleep as well as potential injuries to patients themselves or bed partners [1]. RBD has attracted neurologists' attention extensively over the past decade as they determined it not only a "premotor" PD state but also potentially indicated a more severe outcome among PD patients.

According to International Classification of Sleep Disorders-3, the definite RBD can be diagnosed if all criteria are met as follows: (1) there are repeated episodes of sleep-related vocalization and/or complex motor behaviors detected by a single night of vPSG, which often correlate with simultaneously occurring dream mentation and lead to the frequent report of "acting out one's dreams"; (2) those behaviors documented by PSG occur during REM sleep; (3) there is RWA demonstrated by PSG as defined by the American Academy of Sleep Medicine (AASM) Manual for the Scoring of Sleep and Associated Events; and (4) these disturbances mentioned above cannot be better explained by any other sleep disorders, mental disorder, medication, or substance use. In addition, upon awakening, the individual is typically awake, alert, coherent, and oriented. On occasion, there may be patients who exhibit RBD behaviors during vPSG but do not demonstrate sufficient RWA. They may typically have clinical histories of RBD but do not satisfy the PSG criteria of RBD. In such patients, RBD may be provisionally diagnosed, based on clinical judgment. The same rule also applies when vPSG is not readily available. On other circumstances, medications such as SSRIs may unmask latent RBD with preexisting RWA. These patients can be diagnosed as RBD, but close follow-up is needed [1]. Besides critical diagnostic criteria based on vPSG, several questionnaires as shown in Table 3.1 have been developed and validated to screen and diagnose the presence of RBD.

The differential diagnosis of recurrent dream enactment behavior, as summarized in Table 3.2, includes NREM parasomnia, nocturnal panic attacks, nocturnal seizures, nightmares, nocturnal wandering associated with dementia, and OSAS [6].

In 2009, a 68-year-old man suffering from 2-month violent behavior during sleep was diagnosed with RBD with PSG evidence and simultaneously detected with a lacunar ischemic infarct in the right paramedian pons by MRI [7]. In addition, Nadège et al. once reported that a 40-year-old woman without prior parasomnia developed severe RBD lasting for 2 years after an acute inflammatory rhombencephalitis [8]. In 2010, through a neuroimaging analysis, Ellmore TM et al. showed

Table 3.1 RBD screening questionnaire

Name	Brief description
Mayo Sleep Questionnaire [2]	A 16-item measure screening for the presence of RBD and other sleep disorders [2]
RBD screening questionnaire (RBDSQ) [3]	A 10-item patient self-rating questionnaire covering the clinical features of RBD [3]
RBDQ-HK [4]	A validated 13-item self-reported questionnaire, originally from RBDSQ
RBDSQ-J [5]	Translated into Japanese by Miyamoto et al.

Table 3.2 Differential diagnosis of RBD [6]

Primary disorders of arousal (from NREM sleep)
Confusional arousals
Sleepwalking
Sleep terrors
Secondary arousal disorders
Obstructive sleep apnea syndrome (pseudo RBD)
Sleep-related epilepsy
Psychiatric diseases
Sleep-related dissociative disorder
Panic disorder
Posttraumatic stress syndrome

that bilateral putamen volumes of RBD patients were smaller than the age and gender-matched controls [9]. With the help of neuromelanin-sensitive imaging, Garcia-Lorenzo et al. proved that the reduction of signal intensity in locus coeruleus/subcoeruleus area in patients with both PD and RBD was more marked than PD patients without RBD [10]. These studies together indicated that brainstem is highly associated with RBD. Unfortunately, the specific nuclei or the precise dysfunctional neuronal network is not fully understood yet, which motivates researchers to figure it out.

After Schenck et al. firstly came up with the theory in 1996 that RBD patients could develop a parkinsonian disorder, a few years later, more studies sprang up and verified that RBD could be prodromal symptoms of neurodegenerative diseases, mainly synucleinopathies. Recently researchers have proposed that RBD alone can be a prodromal stage of synucleinopathies. Those with isolated RWA (REM sleep without atonia) exhibited higher risk of developing synuclein-mediated neurodegeneration [11]. Among patients with established PD, the prevalence of polysomnographically defined RBD is 39–46% [12], with 18–52% developing RBD before the onset of PD [1]. PD patients that comorbid with RBD have longer disease duration, higher Hoehn and Yahr stages, more falls, more fluctuations, more psychiatric disorders, and a higher dose of levodopa consumption. The presence of RBD in PD patients is related to a higher amount of REM sleep, more periodic leg movements during sleep, a higher systolic blood pressure change while standing, and worse cognition [1].

Treatment of RBD is composed of two parts. The first one is to reduce the risk of injuries, e.g., moving sharp and edged objects out of harm's way, placing a mattress or cushion on the floor adjacent to the bed, and building a protective barrier on the side of the side. Another one is medication, in which clonazepam and melatonin are commonly used following guidelines established for idiopathic RBD, but data is limited on the efficacy of these drugs in individuals with RBD and a diagnosed PD [12]. Clonazepam could suppress behavioral symptoms and reduce phasic REM muscle activity in patients with RBD but does not restore REM sleep muscle atonia. Side effects of clonazepam include daytime somnolence, cognitive impairment, and aggravation of obstructive sleep apneas (OSA) [13]. In contrast, melatonin has rare side effects, making it a better choice for patients who suffer from OSA or cognitive impairment. Moreover, melatonin decreases the percentage of RWA, indicating that it could have a more direct effect on REM sleep. Recently melatonin receptor agonist Ramelteon has shown positive effect on RBD patients, while large-scale clinical trials are still needed to further assess its efficacy and safety for long-term usage. Pramipexole was also reported using in a few cases to improve clinical manifestations of RBD, with limited data though.

3.1   Case Report

A 72-year-old male was admitted to the Department of Neurology with a chief complaint of sleep disorder in July 5, 2018. He had presented with a 3-year history of repeated dream-enacting behavior for no obvious reason. Following a regular sleep routine of going bed at 8–10 pm and waking up at 5–6 am, he reported having episodes 2 or 3 times and being awaken up by his wife when this occurred, which often interrupted his vivid dream during nighttime every day. The contents of dreams were various but most of them were nightmares concerned with violence, such as fighting with somebody. According to his wife, he often acted out his dream, which included somniloquy, singing, yelling, punching, kicking the door, falling down from bed, and even sometimes provoking lesions to his wife.

The patient had no past history of serious illness, head injury, depression, and use of antidepressants, but had a long history of smoking and drinking alcohol and did not present any relevant family history of disease. Neurological examination was normal and no extrapyramidal manifestation was performed. Laboratory polysomnography (PSG) disclosed sleep efficiency of 57.2%, a sleep latency of 40.5 min, a total time of waking of 207.5 min (32 times). The amount of REM sleep was within normal limits (25% total sleep time), but the deep sleep decreased sharply (% total sleep time). Periodic limb movement had not showed up during sleep. The sleep respiratory pattern matched moderate sleep apnea-hypopnea syndrome, mainly on obstructive sleep apnea-hypopnea syndrome. The AHI measured 19.1 and the lowest oxygen saturation was 75%. One typical characteristic of RBD was atonia during REM sleep, which could be confirmed by PSG (Fig. 3.1). PSG results showed excessive sustained increasement in electromyogram (EMG) amplitude, which presented as twitching in her chin and limbs during REM sleep. The video-PSG result depicted periods of movements of arms and legs and sitting up in bed, lasting 5–20 s each time. Cerebral magnetic resonance imaging showed some lacunar infarction in both frontal lobe and basal ganglia, and mild cerebral atrophy.

According to the ICSD-3 (International Classification of Sleep Disorders, third Edition) and based on the subjective symptoms and the test results, the patient was diagnosed as REM sleep behavior disorder. As for differential diagnosis, RBD should be distinguished from epilepsy, somnambulism, and nightmare. Electroencephalography revealed no epileptic discharges which ruled out the possibility of epilepsy. Polysomnography (PSG) demonstrated that the dream enactment of the patient showed up in rapid eye movement rather than somnambulism

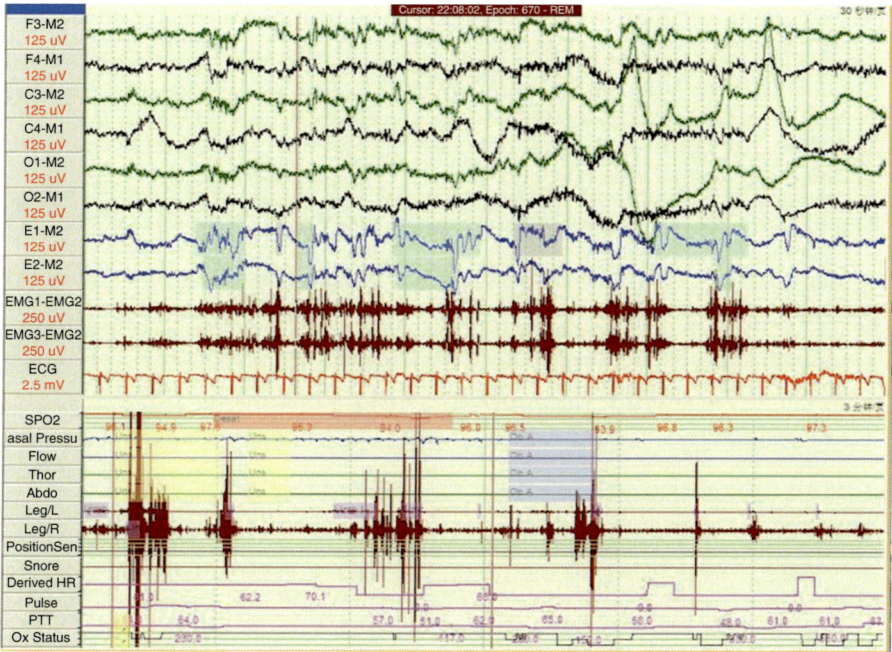

Fig. 3.1 Polysomnographic recording of the patient during REM sleep prior to treatment

occurred in non-rapid eye movement. The DEB (clinical manifestation) pointed out the differentiation between RBD and nightmare.

After a consideration of the therapeutic possibilities, an agreement was made and given a dosage of melatonin, 0.25 mg at bedtime. At the 6-month follow-up visit, the patient noted that his RBD symptoms, the dream enactments, were improved. Except for drug therapy, non-drug therapy is also significant. So, we taught the patient and his wife to creating a safe sleep environment, they removed the bedside cabinet to prevent hitting when he was kicking and fist fighting. And they laid carpet nearby their bed to protect him when he falls down from bed. The patient has not fell down nearly 3 months and the frequency of dream enactment decreased during follow-up. Many clinical studies have shown that RBD may be a precursor syndrome of neurodegenerative disorders (often following a delay of decades). Onset for 4 years, there was no clinical manifestation of neurodegenerative disorders in this patient yet.

References

1. Jiang H, Huang J, Shen Y, Guo S, Wang L, Han C, et al. RBD and neurodegenerative diseases. Mol Neurobiol. 2017;54(4):2997–3006.
2. Boeve BF, Molano JR, Ferman TJ, Smith GE, Lin SC, Bieniek K, et al. Validation of the Mayo sleep questionnaire to screen for REM sleep behavior disorder in an aging and dementia cohort. Sleep Med. 2011;12(5):445–53.

3. Stiasny-Kolster K, Mayer G, Schafer S, Moller JC, Heinzel-Gutenbrunner M, Oertel WH. The REM sleep behavior disorder screening questionnaire—a new diagnostic instrument. Mov Disord. 2007;22(16):2386–93.

4. Li SX, Wing YK, Lam SP, Zhang J, Yu MW, Ho CK, et al. Validation of a new REM sleep behavior disorder questionnaire (RBDQ-HK). Sleep Med. 2010;11(1):43–8.

5. Neikrug AB, Ancoli-Israel S. Diagnostic tools for REM sleep behavior disorder. Sleep Med Rev. 2012;16(5):415–29.

6. Miyamoto T, Miyamoto M, Iwanami M, Kobayashi M, et al. The REM sleep behavior disorder screening questionnaire: validation study of a Japanese version. Sleep Med. 2009;10(10):1151–4.

7. Xi Z, Luning W. REM sleep behavior disorder in a patient with pontine stroke. Sleep Med. 2009;10(1):143–6.

8. Limousin N, Dehais C, Gout O, Heran F, Oudiette D, Arnulf I. A brainstem inflammatory lesion causing REM sleep behavior disorder and sleepwalking (parasomnia overlap disorder). Sleep Med. 2009;10(9):1059–62.

9. Ellmore TM, Hood AJ, Castriotta RJ, Stimming EF, Bick RJ, Schiess MC. Reduced volume of the putamen in REM sleep behavior disorder patients. Parkinsonism Relat Disord. 2010;16(10):645–9.

10. Garcia-Lorenzo D, Longo-Dos Santos C, Ewenczyk C, Leu-Semenescu S, Gallea C, Quattrocchi G, et al. The coeruleus/subcoeruleus complex in rapid eye movement sleep behaviour disorders in Parkinson's disease. Brain. 2013;136:2120–9.

11. Stefani A, Gabelia D, Hogl B, Mitterling T, Mahlknecht P, Stockner H, et al. Long-term follow-up investigation of isolated rapid eye movement sleep without atonia without rapid eye movement sleep behavior disorder: a pilot study. J Clin Sleep Med. 2015;11(11):1273–9.

12. Chahine LM, Amara AW, Videnovic A. A systematic review of the literature on disorders of sleep and wakefulness in Parkinson's disease from 2005 to 2015. Sleep Med Rev. 2017;35:33–50.

13. Gagnon J-F, Postuma RB, Mazza S, Doyon J, Montplaisir J. Rapid-eye-movement sleep behaviour disorder and neurodegenerative diseases. Lancet Neurol. 2006;5(5):424–32.

Insomnia

4

Hua Hu and Chun-Feng Liu

Abstract

We know that sleep problems often occur in patients with PD, such as difficulty in falling asleep, difficulty in maintaining sleep, etc. However, as a key indicator of sleep maintenance insomnia, sleep fragmentation is the most common complaint of sleep.

Risk factors for poor sleep include various physical diseases, female, poor sleep habits, life stress events, etc. Similarly, there are different sleep risk factors in Parkinson's disease. In addition to the motor symptoms associated with PD, such as tremor, stiffness, leg spasm and dystonia, sleep disorders associated with PD are also very common, such as primary sleep disorders, psychophysiological insomnia (independent of the disease), RLS/PLMD, vivid dreams/nightmares and obstructive sleep apnoea (OSA).

Studies have confirmed that the insomnia of PD patients is related to many factors, among which the severe motor and non-motor symptoms are more closely related. Therefore, the quality of life of patients is obviously affected.

Existing evidence supports that the treatment of insomnia with PD is considered insufficient. Controlled-release carbidopa-levodopa, eszopiclone and melatonin 3–5 mg were considered acceptable risks without special monitoring. Although there is limited evidence to support the treatment of other psychiatric disorders that may lead to insomnia, the treatment of insomnia for specific psychiatric symptoms, such as anxiety, depression and hallucinations, is worth considering in clinical trials.

Keywords

Insomnia · Parkinson's disease · Risk factors · Etiologic mechanisms · Clinical manifestations · Diagnostic criteria · Diagnostic instruments · Treatment

H. Hu · C.-F. Liu (✉)
Department of Neurology, The Second Affiliated Hospital of Soochow University, Suzhou, China
e-mail: liuchunfeng@suda.edu.cn

© Springer Nature Singapore Pte Ltd. 2020
C.-F. Liu (ed.), *Sleep Disorders in Parkinson's Disease*,
https://doi.org/10.1007/978-981-15-2481-3_4

25

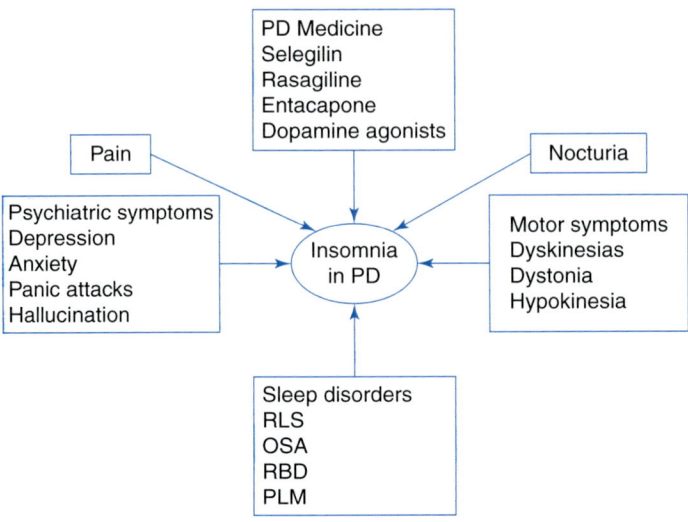

4.1   Introduction

More and more studies show that non-motor symptoms are an important cause of morbidity and disability in patients with Parkinson's disease (PD) [1]. Sleep disturbance is one of the most common non-motor symptoms of Parkinson's disease. Although James Parkinson first mentioned the sleep disorder of Parkinson's disease in his "Essay on the shaking palsy" more than 200 years ago, it is not until recent years that the sleep problem of Parkinson's disease has attracted the attention of scholars. There are many types of sleep disorders in Parkinson's disease, including rapid eye movement (REM) sleep behaviour disorder, obstructive sleep apnoea (OSA), periodic leg movement (PLM), restless leg syndrome (RLS) and insomnia. Insomnia is the most common sleep disorder, defined as difficulty falling asleep at night and daytime energy fatigue. There are many causes of insomnia in PD patients. For patients with more severe symptoms, tremors, stiffness, pain and impaired ability to move in bed often lead to sleep difficulties and insomnia, thus seriously affecting the quality of life of PD patients. PD medicines not only improve patients' motor and non-motor symptoms but also have adverse effects on sleep. In addition, the presence of psychiatric symptoms is also one of the causes of insomnia in PD patients.

4.2   Prevalence of Insomnia in Parkinson's Disease

According to cross-sectional subjective assessment, patients with different types of sleep disorders account for 20–80% of PD patients. In a multi-centre study, 98.6% of the 1072 patients with Parkinson's who were continuously treated or untreated

had non-motor symptoms, and one-third of the subjects complained of sleep disturbances [1]. A national survey conducted in the UK in the 1980s showed that the
majority (76%) of 220 people with PD reported sleep disruption [2]. Similarly,
Ylikoski et al. conducted a study on a large sample of PD patients and found that
81–82% of the subjects had different degrees of sleep disruption [3]. In a study of
66 medicine-naïve patients with PD, 27% of participants self-reported insomnia [4].
It is not clear whether PD patients who did not take medication early were more
likely to suffer from insomnia than those in the control group. A recent study by
Shafazand et al. using semi-structured interviews showed that insomnia was common in Parkinson's patients (46%) [5]. In addition, Chung et al. found that among
128 patients of PD with insomnia, insomnia was associated with levodopa equiva-
lent dose [6]. In summary, compared with the age-matched control group, insomnia
is very common in PD patients. Even after adjusting for depression, anxiety, PD
severity and levodopa dose equivalents, insomnia is still the most common type of
sleep disorder.

4.3   Risk Factors of Insomnia in Parkinson's Disease

There are several potential causes of and/or contributors to insomnia in PD,
including the underlying degeneration of sleep regulatory centres, comorbidity or
the sleep-altering effect of antiparkinsonian medicines [7]. A cross-sectional analysis showed that insomnia was associated with longer disease duration and
occurred more often in females [8]. The association between female sex and
insomnia has also been seen in the general population and hence is not specific to
patients with PD.

Several reports in recent years have suggested that insomnia is associated with
more severe PD motor and non-motor manifestations. Neuropsychiatric symptoms, nocturia, dyskinesias, pain or dystonia [2] as well as intrinsic circadian
rhythm dysregulation [9] could significantly contribute to sleep disruption in PD
patients. Shorter nocturnal sleep times and longer sleep-onset latency seem to be
characteristic of patients with PD that experience an improvement in their motor
symptoms upon awakening in the morning [10]. Autonomic dysfunction (nocturia,
hypokinesia) occurring during sleep leads to disruptive night-time sleep and
insomnia symptoms [11].

Furthermore, depression was a particularly strong risk factor for insomnia. It is
important to realize that insomnia is a characteristic of depression and that therefore
the two features are inherently related. Previous studies in PD showed that insomnia
and depression frequently coexist [12, 13]. Many findings show that female patients
with PD who have longer disease duration and present with depressive symptoms
should be monitored closely for future development of insomnia [14].

Although dopaminergic medicines can improve night-time hypokinesia, dyskinesia and tremor, they might also worsen insomnia. A higher dose of dopamine
agonists was found to be associated with early and frequent awakenings and subjective lack of sleep, but not with sleep initiation or lying awake too long. This implies

that dopamine agonists selectively affect sleep maintenance and not sleep initiation. Therefore, many modifications in timing and dose of dopaminergic medicines or a reduction of the dose of dopamine agonists, or both, should be considered [15].

4.4 Etiologic Mechanisms of Insomnia in Parkinson's Disease

As we all know, sleep is one of the most important parts, and once insomnia has a considerable impact on health. Insomnia in PD patients has different degrees of influence on their own motor symptoms and non-motor symptoms. And they influence each other. The mechanisms behind these observations, however, remain unclear.

The possible underlying aetiology of insomnia in this population is thought to be multifactorial and in part due to intrinsic abnormalities in central nervous system structure and function account for sleep problems. Degeneration of brainstem/thalamocortical regions involved in the sleep-wake cycle likely contribute to insomnia in PD, as do abnormalities in the suprachiasmatic nucleus and other hypothalamic structures. Another important consideration is the effect of dopamine medications, especially at high doses, and the duration of action of dopamine agonists is also an important aspect of treating insomnia [7]. The relationship between dopamine agonists and insomnia is complex and controversial. Although certain dopamine agonists may worsen insomnia, some researchers have shown that they improve sleep quality in PD patients [16].

Dopamine agonists may affect sleep in PD patients in different ways. First, dopamine treatment may increase the risk of hallucinations, which in turn could cause nocturnal sleep disturbances [17]. Second, dopamine plays an important role in regulating sleep-wake rhythms [7]. Dopamine agonists have a bipolar effect on sleep arousal, which can be attributed to stimulation of D2 receptors. At low doses, they reduce arousal and enhance sleep, while at higher doses they have the opposite effect, causing insomnia [16]. Finally, the fluctuation of motor symptoms is closely related to the aggravation of insomnia symptoms over time, and the complications of levodopa treatment usually increase the prevalence and severity of insomnia as PD progresses. At present, there are several new effective strategies for the motor fluctuation, which may have a beneficial effect on the insomnia of PD patients [18].

4.5 Clinical Manifestations of Insomnia in Parkinson's Disease

Sleep is the most important part of restoring one's strength and energy. Most adults sleep 5–8 h. Sleep requirements change with physical activity, age, environment, pregnancy, and other physical illnesses. Chronic sleep deprivation can lead to reduced productivity. Excessive fatigue not only affects our daily life and work but also has a huge impact on our mind and body, increasing the risk of physical illness and accidents.

The prodromal phase of PD is usually characterized by sleep changes [19, 20]. In addition to the early stages of the disease, changes in sleep may also trigger subsequent neurodegenerative changes [21, 22]. Therefore, it is worth noting that in the prodrome of PD, patients may first develop sleep disorders before any obvious symptoms such as movement or mood appear.

A lot of research on sleep disorders in PD has focused on REM sleep behaviour disorder (RBD), a common sleep disorder that may predate movement symptoms in PD [23]. In fact, insomnia is the most common symptom in PD, and past studies have shown that even mild untreated PD patients have significantly reduced total sleep time compared with healthy age-matched controls [24]. The American Academy of Sleep Medicine defines insomnia as difficulty falling asleep, difficulty maintaining sleep, early awakening, poor overall sleep quality, and poor daytime energy. In PD, persistent sleep disruption and early awakening are the most common symptoms, and the onset of sleep is often unaffected [25].

Sleep plays an important role in the maintenance and consolidation of memory. Synaptic remodeling influences initial learning and long-term memory consolidation [26, 27]. Chronic sleep deprivation increases cognitive decline and memory loss, leading to the risk of dementia [28]. Chronic sleep deprivation can lead to depression, anxiety and other psychological problems [29, 30]. K. Zhu et al. found a two-way relationship between these symptoms, that is, depressive symptoms may lead to insomnia, which in turn may lead to the development of PD depression. In summary, poor sleep quality in Parkinson's disease is particularly associated with significant cognitive deficits and mood disorders, as well as dementia and depression [31].

In addition, Parkinson's insomnia affects the quality of work and life the next day and can lead to excessive daytime sleepiness (EDS) and energy fatigue. Insomnia affects blood pressure, oxygen and carbon dioxide levels and is therefore a common risk factor for cardiovascular and cerebrovascular diseases. On the other hand, insomnia can impair social and work functions. If the subjects drive after insomnia, the risk of accidents will be significantly increased.

4.6 Diagnostic Criteria and Instruments of Insomnia in Parkinson's Disease

Insomnia is a common disease in patients with movement disorder, which has been widely studied in PD. At present, the diagnosis depends on the patient's chief complaint, rather than a specific amount of sleep or other objective sleep monitoring.

The diagnostic criteria of "insomnia" include (1) difficulty in falling asleep, easy to wake up at night or early wake up; (2) despite adequate sleeping environment; and (3) lack of sleep that causes daytime symptoms such as attention problems, memory problems or energy fatigue.

Insomnia can be divided into primary insomnia and secondary insomnia. However, it is often difficult to determine the causes or factors associated with insomnia. Therefore, "comorbidity" is preferable to "secondary".

Questionnaires are often used to objectively assess unobservable symptoms and signs (i.e., sleep problems), determine severity, impact on quality of life and assess changes in sleep before and after medication intervention. Currently, a variety of questionnaire-based tools have been used to investigate insomnia in patients with PD. These questionnaires include the Pittsburgh sleep quality index (PSQI), Parkinson's sleep scale (PDSS—version 1 and version 2), Parkinson's disease, the sleep scale (SCOPA sleep), the motor symptoms scale (NMSquest), sleep quality index (PSQI), the sleep quality index (PSQI), the sleep quality index (PDSS—version 1 and version 2), sleep quality index (SCPA), sleep non-motor symptoms scale (NMSS), the movement disorder society unified Parkinson disease rating scale (MDS—UPDRS), Athens insomnia scale (AIS) and Epworth Sleepiness Score (ESS).

Pittsburgh sleep quality index (PSQI) is a general 19-item self-rating scale designed to measure overall sleep problems. Seven component scores (subjective sleep quality, sleep latency, sleep duration, habitual sleep efficiency, sleep disorders, use of sleep medicines and daytime dysfunction) were synthesized from 19 project groups. Items score from 0 to 3 (no difficulty to severe difficulty). On a scale of 0–21, the higher the score, the more serious it is. The scale has been widely used for primary insomnia, dementia, depression and anxiety as well as for motor disorders such as PD. After interventions such as deep brain stimulation [32] and medicine therapy, it shows sensitivity to changes in PD populations [33].

Parkinson's disease sleep scale (PDSS) is a validated visual simulation scale designed to measure sleep complaints in patients with PD. PDSS covered 15 symptoms commonly associated with PD sleep disorders, including overall sleep quality, sleep onset and maintenance insomnia, nocturnal restlessness, nocturnal psychosis, nocturia, nocturnal motor symptoms, refreshing sleep, and daytime sleepiness. Each item was scored on a 0–10 visual analogy scale (VAS) with an overall score of 0–150. Scores higher than 82 indicate acceptable sensitivity and specificity to nocturnal sleep problems [34]. PDSS has been widely used in PD and has been studied in the dystonia cohort [35].

Parkinson's Disease Sleep Scale-2 (PDSS-2) is the second version of PDSS, with two major differences from the first (see above): items were rated on the Likert scale (from 0, never, to 4, very frequent), and all 15 items were assessed for nocturnal sleep problems. The total score is between 0 and 60, with higher scores indicating higher severity. Score higher than or equal to 15, can distinguish "bad" sleepers and "good" sleepers, diagnostic accuracy acceptable [36].

Scales for Outcomes in PD-Sleep (SCOPA-Sleep) is a PD-specific scale that includes 12 items to measure sleep quality, nocturnal sleep disorders and daytime sleepiness. The night-time sleep subscale included five items for insomnia, multiple awakenings, sleep efficiency and duration and one item for overall sleep quality. Patients with a score of 7 showed good sensitivity and satisfactory specificity [37].

Multidomain PD-Specific Scales, such as the Non-Motor Symptoms Questionnaire (NMSQuest), the Non-Motor Symptom Scale (NMSS), and the Movement Disorder Society Unified Parkinson Disease Rating Scale (MDS UPDRS), part 1B, include a single item for evaluating insomnia. The NMSS is a score-based scale that assessed the severity and frequency of 30 different non-motor symptoms of PD over the past month. Item 5 was used to assess difficulty in falling asleep or staying asleep in other low motivation disorders such as MSA and PSP [38]. NMSQuest and the MDS UPDRS part 1B are self-assessment questionnaires. Question 23 on NMSquest assessed difficulty falling asleep or maintaining sleep, while item 1.7 of the MDS UPDRS asked patients to rate overall sleep quality over the past seven days. Another disease-specific scale for another form of PD, the Progressive Supranuclear Palsy Rating Scale, which includes difficulty sleeping during daily activities, is rated from 0 to 4.

Athens insomnia scale (AIS) was used to evaluate the symptoms of insomnia. The first five were related to sleep induction, nocturnal arousal, final arousal, total sleep duration and sleep quality. The last three are related to health, functional ability and sleepiness [39].

Epworth Sleepiness Score (ESS) was used to measure subjective daytime sleepiness [40]. The overall score ranges from 0 to 24, with higher scores indicating easier sleep under different conditions.

4.7   Treatment of Insomnia in Parkinson's Disease

Poor sleep is common in PD and has been reported to negatively affect patients' quality of life. However, as for the treatment, sleep symptoms have not received enough clinical attention. As a chronic disease, treatment is still difficult to determine, and identifying and treating sleep disorders and subsequently improving quality of life is an important therapeutic goal. Treatment includes sleep hygiene and consideration of whether to use hypnotics or sedative antidepressants.

Sleep Hygiene
(a) Keep regular sleep and wakefulness habits. This strengthens circadian rhythms. Set an alarm and get up at the same time every morning, no matter how much sleep you get at night.
(b) Exercise regularly, but not too vigorously.
(c) A light meal or hot milk drink can induce sleep. Don't eat too much at night.
(d) Don't drink coffee or tea before bed. Don't try to drink alcohol to improve your sleep.
(e) Create a conductive environment—a cool, quiet room and a solid, comfortable but not rigid bed.
(f) Sleep regulation—the principle is to associate bedroom stimulation with sleep. Stay in bed only when you are sleepy. The bed can only be used for sleeping, not for eating, writing or watching TV. If you can't fall asleep after 30 min at night, get out of your bedroom and read or write until you're tired, then go back to your bedroom and sleep. If sleep is not easy, get up again and repeat the procedure.
(g) Muscle relaxation—progressive muscle relaxation consists of a series of movements that contract and relax different muscle groups. Focus on the feeling of relaxation, which leads to sleep (see Sect. 4.9).

Light Therapy

Altemenko and Levine used bright light therapy (LT) to treat PD-related depression. This approach is based on evidence that the strategic use of LT has a beneficial effect on seasonal affective disorder. It not only improved the depression associated with PD but also improved some aspects of motor function after only a few PD exposures. In addition, it has been found to be effective for insomnia.

Previous studies have reported that light use in treatment regimens is beneficial for PD patients, but these studies have been limited to short-term and morning LT patients [9]. The LT's standard solution is a 2500 lx white fluorescent lamp that emits 20 min of light each morning from sleep. Given that PD is a chronic and progressive disease, Jessica et al. demonstrated that applying LT before retirement improves sleep quality and reduces the incidence of nocturnal activity [41]. Clearly, the structure of sleep changes in the first few days after the onset of sleep. Not only does the total number of hours of sleep begin to increase, but the number and duration of awakenings usually decrease as patients fall asleep more easily. Improvements can be maintained as long as they continue over a period of 4–6 years. These results demonstrate the value of long-term use of non-invasive techniques (e.g., LT) to treat sleep disorders in PD patients and justify further long-term controlled trials of efficacy. In addition, hypnotics and time-therapy medicines are often used in combination with PD patients to achieve better therapeutic effects.

It involves daily exposure to bright artificial light, usually using fluorescent light boxes. New light-processing devices include those that use light-emitting diodes (LEDs). They have the advantages of long life, portability and different wavelength. It works by transferring retinol from the eyes to the brain through the hypothalamus nervous system. The therapeutic effect mainly involves circadian rhythm regulation

Pharmacotherapy

Not everyone who has trouble sleeping should take sleeping pills. Benzodiazepines are the most commonly used hypnotics, but can cause daytime sleepiness and reduced productivity. Sudden withdrawal after long-term use can cause rebound insomnia. Before prescribing hypnotics, attention should be paid to (1) the relationship between insomnia and mood, whether there is anxiety, depression and other symptoms; (2) the nature of sleep disturbance, whether there is difficulty in falling asleep, frequent awakening or early awakening; (3) whether patients need to be alert at all times during the day work; and (4) combined with alcohol, anti-histamines, antidepressants, antipsychotics, anti-anxiety medicines and narcotic analgesics, can improve the sedative effect (Table 4.1).

Current guidelines for insomnia recommend short-intermediate-acting benzodiazepine receptor agonists, such as eszopiclone. Many single trial results show that eszopiclone, doxepin and melatonin can effectively treat insomnia in PD patients. Melatonin (3 mg 1–2 h before bedtime) not only contributes to the treatment of insomnia in some patients but also can effectively treat RBD.

Among PD medications, selegiline, which is metabolized to amphetamine, is one of the most likely to cause insomnia. Sleep disorders are more common in patients

Table 4.1 Pharmacokinetics of the common hypnotics

	Dosage (mg)	Duration of action (h)
Diazepam	5	12~14
Flurazepam	15	8~12
Nitrazepam	5	8
Lormetazepam	1	6
Zopiclone	7.5	6
Zolpidem	10	6
Eszopiclone	3	6

taking dopaminergic agonists (DAS). However, those who took DAS continuously had less sleep disturbance than those who took DAS newly. People who stop DAS have increased sleep problems. In clinical practice, controlled release (CR) levodopa is often used in the treatment of nocturnal akinesia in PD patients. Dopamine agonists, rotigotine, ropinirole and pramipexole also improve sleep quality in PD patients.

In addition, a large randomized placebo-controlled trial showed that pimavanserin may improve sleep at night in PD patients with psychosis [42], while norteterine (not paroxetine) improved sleep quality in patients with depression [33]. In an open-label study, tolcapone improved sleep quality [43], while a retrospective study showed clozapine improved sleep disorders in PD patients [44]. If depression is a factor, use sedatives or daytime antidepressants, such as selective serotonin reuptake inhibitors (SSRI) and night-time hypnotics.

Transcranial Magnetic Stimulation
Transcranial magnetic stimulation (TMS) is a tool for non-invasive magnetic stimulation of the cerebral cortex. After 5 Hz stimulation of the primary motor cortex for 10 consecutive days, PD patients reported improved sleep on the self-assessment scale [45].

In 2009, Karin et al. characterized sleep in patients with PD and examined the effects of rTMS (repeated transcranial magnetic stimulation) in combination with actigraphy and pressure-sensitive pads [46].

The mixed model regression analysis showed that compared with the healthy control group, PD patients had shorter sleep time, more debris, lower sleep efficiency and longer waking time at night. rTMS in the parietal lobe, not the motor cortex, improved sleep fragmentation and sleep efficiency and reduced the average duration of nocturnal arousal. There was no effect on motor symptoms.

Psychotherapy
Cognitive therapy (CT) is the most validated psychosocial treatment in psychiatry, with numerous RCTs and meta-analyses showing evidence for efficacy in insomnia. In CT, automatic negative thoughts associated with sleep are believed to underlie the insomnia feelings. CT seeks to systematically identify these thinking patterns and then rationally challenge them. Techniques of CT include keeping track of automatic thoughts, assessing the affect associated with them, and then reassessing feeling after an intervention such as a rational challenge. Homework assignment and review is an integral part of CT.

Behaviour therapy (BT), or behavioural activation, is also a widely validated treatment for insomnia. This sets up a "vicious circle" in which reduced activity leads to more inertia that further limits activity. In BT, the inertia and reduction in goal-directed activities are addressed to facilitate new learned behaviours. Negative behaviours are used, including increasing the number of pleasurable activities, tracking moods, relaxation exercises and social skills training.

Cognitive therapy and behaviour therapy are often used together as *cognitive behavioural therapy (CBT)*, especially in community practice. CBT is also widely used in other psychiatric and medical conditions. CBT has been found to yield therapeutic efficacy in primary insomnia and should be considered for PD patients as well. Preliminary studies suggest that these may be effective interventions for insomnia, which would make CBT much more accessible to PD patients.

Group psychotherapies have been shown to be effective for insomnia. The advantages of group psychotherapy are that patients have more sources for support and encouragement, a group offers the opportunity to practice interpersonal techniques and receive feedback, and groups may be more cost-effective. Some of potential disadvantages of group psychotherapy are that often patients will prefer individual therapy, individual therapy may be more effective, and it is more difficult to engage patients and schedule appointments for groups.

4.8 Concluding Remarks

Insomnia is common in PD and there are several underlying causes and/or causes of PD insomnia. Depressive symptoms, motor fluctuations, longer disease duration, females, and the use of larger doses of dopamine agonists are all associated with insomnia. Paying attention to these aspects may help better understand the symptoms of insomnia in PD. However, the underlying causes of insomnia in PD are thought to be multifactorial, and the exact mechanism is still unclear. The American Academy of Sleep Medicine defines insomnia as difficulty falling asleep, wakefulness, early wakefulness and daytime energy fatigue. Currently, diagnosis is largely self-reported. Multiple questionnaire scales can be used to assess insomnia in PD. Insomnia in PD has a significant effect on the quality of life of patients. Treatment of insomnia in PD includes sleep hygiene and the use of different hypnotic medicines. Non-pharmacological therapies such as light therapy, rTMS and psychotherapy have also been shown to be effective. Hopefully, more research will be done to explore more effective treatments for insomnia in PD.

4.9 Case Report

4.9.1 Clinical Feature

Male, 68 years old, has a 5-year history of PD. The patient initially presented with resting tremor and bradykinesia of the left upper limb. Symptoms subsequently appear in the left and right lower extremities. In 2013, he was diagnosed with PD. He took anti-Parkinson's medicines and his symptoms improved significantly. He has a long history of insomnia, which worsened dramatically in the last year, forcing him to retire from security. He calls himself a "late sleeper". He does not believe that his motor symptoms, nocturia and night-time pain are the causes of his night-time sleep problems. He denied history of restless legs syndrome (RLS). He complained of snoring, but refused continuous positive airway pressure to treat mild obstructive sleep apnoea (OSA). However, he had interest in daily life and had not obvious depressive and anxious symptoms, willing to communicate with others. At admission to our hospital, he had a normal diet, extremely inadequate sleep (about 2–3 h per day on average). It took a long time to fall asleep. Even if he fall asleep, it's very easy to wake up. He reported waking up feeling unrefreshed and had to take 2 h of rest at noon; otherwise he would be dizzy in the whole afternoon. Memory is not as good as before. Because of sleep disorder, he did not go out to work and could only do some chores for his wife.

4.9.2   Past History, Family History and Personal History

He had no previous history of other medical illness. His mother had suffered from PD, with an initial onset around 57 years old. Physical sign, growing development, major organ index, blood routine examination and blood biochemical criterion were normal. After graduating from junior high school, he had been working as worker or security guard. He has a gentle temper and a little introverted. As his wife had said, he had no vices.

4.9.3   Physical Examination

The neurological examination on admission found slightly clumsy speech patterns, masking face and increased muscle tone and hyperactive deep tendon reflexes in the limbs. All other parameters of the neurological examination were within normal limits.

4.9.4   Imageological Examination and Scale Assessment

In addition, nonenhanced cranial magnetic resonance imaging (MRI) scan and Doppler ultrasound of the head and neck did not identify any evidence of pathology.

Emotion symptoms and cognitive function were assessed clinically using several rating systems, including the Unified Parkinson's Disease Rating Scale (UPDRS I, 20 scores; UPDRS II, 8 scores), Hamilton Depression Rating Scale (HAMD, 6 scores), Hamilton Anxiety Rating Scale (HAM-A, 5 scores) and Mini-Mental State Examination (MMSE, 27 scores).

4.9.5 Polysomnography (PSG)

On one night monitored by the sleep centre polysomnography (PSG), his apnoea-hypopnea index (AHI) is 7.8 and REM-AHI is 15.4, and he has no periodic limb movements (PLMS), and arousal index (AI; number of EEG arousals per hour) was 11.3.

4.9.6 Diagnosis

The patient was diagnosed with sleep maintenance insomnia, delayed sleep-wake phase disorder and mild obstructive sleep apnoea (OSA).

4.9.7 Treatment

4.9.7.1 Sleep Hygiene

In the management of insomnia, it is important to invest some time to explain to the patient the overall treatment plan. Sleep hygiene is as necessary as medications. Gaining the trust and confidence of the patient will prevent unnecessary abuse of benzodiazepines or other hypnotics.

It's important to maintain regular sleep habits. Set an alarm clock and get up at the same time every morning, no matter how long you sleep at night. Exercise regularly, but not too vigorously. Don't drink too much liquid at night, as a full bladder can disrupt sleep. A light dinner or hot milk helps you sleep. Don't drink coffee or tea before bed. Don't try to drink to improve your sleep. Create a comfortable environment—a cool, quiet room and a solid, comfortable but not hard bed. Have a sleep regulation—the principle is to associate the bedroom with sleep, not with the frustration of insomnia. The bed should only be used for sleeping, not for eating, writing or watching TV. If you can't fall asleep within 30 min, get out of the bedroom and go back to sleep until you're tired. If you still can't sleep, you need to get up again and repeat the procedure.

4.9.7.2 Muscle Relaxation Therapy

Progressive muscle relaxation consists of a series of exercises that contract and relax different muscle groups. Focus on the feeling of muscle relaxation, and this slow relaxation helps improve sleep.

First, choose a quiet room. Close the curtains and sit in a comfortable armchair or lie in bed. Close your eyes. Now breathe in slowly, exhale slowly and focus on your right hand and arm. Squeeze your right hand, squeeze hard and then slowly open your hand, releasing the strength of your right hand and arm. Then turn your attention to your left arm. Make a fist out of your hand, clench it and slowly release the tension. Push your shoulders up to your ears, straining the muscles in your neck and back of your shoulders. Keep it up until you can really feel the tension. Slowly relax the tension in your shoulders. Let all the tension flow out of the shoulder muscles. Now let's look at the facial muscles. Bite a little, not too hard, so as not to hurt the teeth and cause some tension in the jaw. Now relax slowly. Close your eyes and hold until you can really feel the tension in your eye muscles, and then slowly relax. Keep all your toes pointed at you, while pulling all the muscles in your legs tight. Tighten all the muscles in your legs. Now relax slowly. Take another deep breath. Breathe in and out slowly, paying attention to the feeling of relaxation. You will feel very fresh.

4.9.7.3 Medication

It is best to plan to use the medicine two to four times a week. This ensures that patients get at least a few good nights of sleep a week, which reduces anxiety and reduces the risk of tolerance. In general, it is recommended to prescribe the medication up to 20 times, if possible, during any 1 month period.

The patient was given 1.5 mg of eszopiclone before bed. The medicine is commonly used for primary insomnia, chronic insomnia, temporary insomnia and residual insomnia after antidepressant treatment. The medicine may selectively bind to a subtype of the benzodiazepine receptor, the alpha 1 subtype, and may enhance GABA inhibition, providing sedative and hypnotic effects more selectively than other effects of GABA. Eszopiclone has been shown to have little or no tolerance or dependence. After taking this medicine, the sleep quality was improved, and the influence on the total waking time and the number of waking times at night was reduced. This medicine may even be safe to consider in PD patients who require treatment with a hypnotic. And rebound insomnia does not appear to this patient.

References

1. Barone P, Antonini A, Colosimo C, et al. The PRIAMO study: a multicenter assessment of nonmotor symptoms and their impact on quality of life in Parkinson's disease. Mov Disord. 2009;24:1641–9.
2. Abdulla AJJ, Pearce VR. The nighttime problems of Parkinson's disease. Clin Neuropharmacol. 1988;11:512–9.
3. Ylikoski A, Martikainen K, Partinen M. Parkinson's disease and restless legs syndrome. Eur Neurol. 2015;73:212.
4. Erro R, Santangelo G, Picillo M, et al. Link between non-motor symptoms and cognitive dysfunctions in de novo, drug-naive PD patients. J Neurol. 2012;259:1808–13.
5. Shafazand S, Wallace DM, Arheart KL, et al. Insomnia, sleep quality, and quality of life in mild to moderate Parkinson's disease. Ann Am Thorac Soc. 2017;14:412–9.
6. Seockhoon C, Bohnen NI, Albin RL, Frey KA, Müller MLTM, Chervin RD. Insomnia and sleepiness in Parkinson disease: associations with symptoms and comorbidities. J Clin Sleep Med. 2013;9:1131–7.
7. Chahine LM, Daley J, Horn S, et al. Association between dopaminergic medications and nocturnal sleep in early-stage Parkinson's disease. Parkinsonism Relat Disord. 2013;19:859–63.
8. Zhu K, van Hilten JJ, Marinus J. The course of insomnia in Parkinson's disease. Parkinsonism Relat Disord. 2016;33:51–7.
9. Videnovic A, Willis GL. Circadian system—a novel diagnostic and therapeutic target in Parkinson's disease? Mov Disord. 2016;31:260–9.

10. Sherif E, Valko PO, Overeem S, Baumann CR. Sleep benefit in Parkinson's disease is associated with short sleep times. Parkinsonism Relat Disord. 2014;20:116–8.
11. Dhawan V, Healy DG, Pal S, Chaudhuri KR. Sleep-related problems of Parkinson's disease. Age Ageing. 2006;35:220–8.
12. Duncan GW, Khoo TK, Yarnall AJ, et al. Health-related quality of life in early Parkinson's disease: the impact of nonmotor symptoms. Mov Disord. 2014;29:195–202.
13. Verbaan D, Van-Rooden SM, Visser M, Marinus J, Van-Hilten J. Nighttime sleep problems and daytime sleepiness in Parkinson's disease. Mov Disord. 2010;23:35–41.
14. Marinus J, Zhu K, Marras C, Aarsland D, van Hilten J. Risk factors for non-motor symptoms in Parkinson's disease. Lancet Neurol. 2018;17:559–68.
15. Schrempf W, Brandt MD, Storch A, Reichmann H. Sleep disorders in Parkinson's disease. J Parkinsons Dis. 2014;4:211–21.
16. Rodrigues TM, Castro CA, Ferreira JJ. Pharmacological interventions for daytime sleepiness and sleep disorders in Parkinson's disease: systematic review and meta-analysis. Parkinsonism Relat Disord. 2016;27:25–34.
17. MD PM-MP, Fabrizio Stocchi MD, FRCP KSM, et al. Prevalence of nonmotor symptoms in Parkinson's disease in an international setting; study using nonmotor symptoms questionnaire in 545 patients. Mov Disord. 2010;22:1623–9.
18. Rascol O, Perez-Lloret S, Ferreira JJ. New treatments for levodopa-induced motor complications. Mov Disord. 2015;30:1451–60.
19. Tekriwal A, Kern DS, Tsai J, et al. REM sleep behaviour disorder: prodromal and mechanistic insights for Parkinson's disease. J Neurol Neurosurg Psychiatry. 2016;88:445–51. https://doi.org/10.1136/jnnp-2016-314471.
20. Iranzo A, Tolosa E, Gelpi E, et al. Neurodegenerative disease status and post-mortem pathology in idiopathic rapid-eye-movement sleep behaviour disorder: an observational cohort study. Lancet Neurol. 2013;12:443–53.
21. Abbott SM, Videnovic A. Chronic sleep disturbance and neural injury: links to neurodegenerative disease. Nat Sci Sleep. 2016;8:55.
22. Musiek ES, Holtzman DM. Mechanisms linking circadian clocks, sleep, and neurodegeneration. Science. 2016;354:1004–8.
23. Wallace DM, Shafazand S, Carvalho DZ, et al. Sleep-related falling out of bed in Parkinson's disease. J Clin Neurol. 2012;8:51.
24. Garcia-Borreguero D, Larrosa O, Bravo M. Parkinson's disease and sleep. Sleep Med Rev. 2003;7:115–29.
25. Elise Tandberg MD, Larsen JP, Karen Karlsen MD. A community-based study of sleep disorders in patients with Parkinson's disease. Mov Disord. 1998;13:895.
26. Abel T, Havekes R, Saletin JM, Walker MP. Sleep, plasticity and memory from molecules to whole-brain networks. Curr Biol. 2013;23:R774–88.
27. Patti CL, Zanin KA, Sanday L, et al. Effects of sleep deprivation on memory in mice: role of state-dependent learning. Sleep. 2010;33:1669.
28. Yaffe K, Falvey CM, Hoang T. Connections between sleep and cognition in older adults. Lancet Neurol. 2014;13:1017–28.
29. Van Someren EJ, Cirelli C, Dijk DJ, Van CE, Schwartz S, Chee MW. Disrupted sleep: from molecules to cognition. J Neurosci. 2015;35:13889–95.
30. Luik AI, Zuurbier LA, Direk N, Hofman A, Van Someren EJ, Tiemeier H. 24-hour activity rhythm and sleep disturbances in depression and anxiety: a population-based study of middle-aged and older persons. Depress Anxiety. 2015;32:684–92.
31. Postuma RB, Aarsland D, Barone P, et al. Identifying prodromal Parkinson's disease: premotor disorders in Parkinson's disease. Mov Disord. 2012;27:617–26.
32. Iranzo A, Valldeoriola F, Santamaría J, Tolosa E, Rumià J. Sleep symptoms and polysomnographic architecture in advanced Parkinson's disease after chronic bilateral subthalamic stimulation. J Neurol Neurosurg Psychiatry. 2002;72:661–4.
33. Menza M, Dobkin RD, Marin H, et al. A controlled trial of antidepressants in patients with Parkinson disease and depression. Curr Neurol Neurosci Rep. 2009;9:263.

34. Martinez-Martin P, Visser MB, Rodriguez-Blazquez C, Marinus J, Chaudhuri K, Van-Hilten J. SCOPA-sleep and PDSS: two scales for assessment of sleep disorder in Parkinson's disease. Mov Disord. 2010;23:1681–8.

35. Shukla AW, Brown R, Heese K, et al. High rates of fatigue and sleep disturbances in dystonia. Int J Neurosci. 2015;126:928–35.

36. Trenkwalder C, Kohnen R, Högl B, et al. Parkinson's disease sleep scale—validation of the revised version PDSS-2. Mov Disord. 2011;26:644–52.

37. Marinus J, Visser M, van Hilten JJ, Lammers GJ, Stiggelbout AM. Assessment of sleep and sleepiness in Parkinson disease. Sleep. 2003;26:1049.

38. Lee CN, Kim M, Lee HM, Jang JW. The interrelationship between non-motor symptoms in atypical Parkinsonism. J Neurol Sci. 2013;327:15–21.

39. Soldatos CR, Dikeos DG, Paparrigopoulos TJ. The diagnostic validity of the Athens insomnia scale. J Psychosom Res. 2003;55:263–7.

40. Johns MW. A new method for measuring daytime sleepiness: the Epworth sleepiness scale. Sleep. 1991;14:540.

41. Martino JT, Freelance CB, Willis GL. The effect of light exposure on insomnia and nocturnal movement in Parkinson's disease: an open label, retrospective, longitudinal study. Sleep Med. 2018;44:24–31.

42. Cummings J, Isaacson S, Mills R, et al. Pimavanserin for patients with Parkinson's disease psychosis: a randomised, placebo-controlled phase 3 trial. Lancet. 2014;383:533–40.

43. Ebersbach G, Hahn K, Lorrain M, Storch A. Tolcapone improves sleep in patients with advanced Parkinson's disease (PD). Arch Gerontol Geriatr. 2010;51:e125–8.

44. Trosch RM, Friedman JH, Lannon MC, et al. Clozapine use in Parkinson's disease: a retrospective analysis of a large multicentered clinical experience. Mov Disord. 1998;13:377–82.

45. Khedr EM, Farweez HM, Islam H. Therapeutic effect of repetitive transcranial magnetic stimulation on motor function in Parkinson's disease patients. Eur J Neurol. 2015;10:567–72.

46. van Dijk KD, Møst EI, Van Someren EJ, Berendse HW, Yd VDW. Beneficial effect of transcranial magnetic stimulation on sleep in Parkinson's disease &dagger. Mov Disord. 2010;24:878–84.

Nocturia

5

Xiao-jing Gu, Bei Cao, and Hui-fang Shang

Abstracts

Nocturia is a used to define the condition in which the individual has to wake at night one or more times to void, which was frequently complained by more than 60% patients with PD. The pathophysiological mechanisms causing nocturia in patients with PD remains poorly understood. Decreased bladder capacity, nocturnal polyuria, circadian rhythm disturbances, some living habits, and comorbidities are thought to be contributing to nocturia in patients with PD. For the diagnosis of nocturia in PD, detailed medical history and urinary diary are needed. More importantly, ultrasonography and urodynamics also help. Nocturia in PD patients has detrimental impact on patients' quality of life; therefore, multiple disciplinary managements are needed to ameliorate symptoms. These include pharmacological management, lifestyle change, and urology referral when necessary. Medication for treating nocturia in PD includes dopamine agonists, antimuscarinic agents, desmopressin, and melatonin. Moreover, intermittent catheterization, botulinum toxin injection, deep brain stimulation, and neuromodulation are also efficient for the management of nocturia in PD.

Keywords

Nocturia · Reduced bladder capacity · Nocturnal polyuria · Circadian rhythm disturbances · Medication · Intermittent catheterization · Botulinum toxin injection · Deep brain stimulation · Neuromodulation

X.-j. Gu · B. Cao · H.-f. Shang (✉)
Department of Neurology, West China Hospital, Sichuan University, Chengdu, Sichuan, China

© Springer Nature Singapore Pte Ltd. 2020
C.-F. Liu (ed.), *Sleep Disorders in Parkinson's Disease*,
https://doi.org/10.1007/978-981-15-2481-3_5

Case

A 68-year-old female patient with 8 years' duration of Parkinson's disease came to the outpatient clinic and complained that she had to wake to void 2–3 times per night since 2 years ago. The patient was asked to keep a 72-h bladder diary, and the results showed an excessive production of urine at night (>33% of the entire day production). Notably, the patient had a history of coronary heart disease for 20 years, and she consumed two cups of tea every afternoon. The laboratory data showed normal urea and creatinine concentrations and routine urine test showed no infection as well. On urinary ultrasonography, her kidneys and bladder were normal, but there was a post-void residual urine volume of 150 ml.

The patient had nocturnal polyuria and incomplete bladder emptying. She met her doctor and was advised to stop tea consumption and was encouraged to exercise and elevate her lower limb above the heart level in the afternoon. The patient was reevaluated 3 months later. Her symptoms were alleviated a little but still waked up one to two times to void every night, which influenced greatly on her quality of life. Therefore, she was prescribed with solifenacin 5 mg and was gradually increased to 10 mg until she can tolerate. Three months later, the patient revisited and reported a remarkable improvement in incomplete voiding. However, she still had complaint of nocturia, even though she was using the catheter just before going to bed. Therefore, she was prescribed with rotigotine transdermal patch at 2 mg per 24 h. At review after 6 months, she was doing well.

5.1 Introduction

Parkinson's disease (PD) is a progressive neurodegenerative disease characterized by cardinal motor symptoms including bradykinesia, rigidity, tremor, and postural instability. Furthermore, non-motor symptoms (NMSs) have also been recognized in many patients. Lower urinary tract (LUT) symptoms, for example, frequent micturition, urgency of urination, urinary incontinence, and nocturia, are common NMSs presenting in 38–71% of patients with PD [1].

LUT symptoms can be classified into two categories: voiding problems (hesitancy, straining to void, interrupted stream retention of urine) and storage problems (urgency, frequency, incontinence, and nocturia) [2].

Among these LUT symptoms, nocturia, defined as "the complaint that the individual has to wake at night one or more times to void" [3], was one of the most frequent complaint in more than 60% patients with PD [1]. Studies have found that older age, male patients, more severe disease stage, and other NMSs such as anxiety were correlated with an increased risk for nocturia in patients with PD [4, 5]. Moreover, nocturia has largely affected patients' quality of sleep and life and increased the burden for caregiver [6] and is also associated with sleep disturbance, falls, and hip fracture in patients with PD [7]. Therefore, it is of great importance and necessity to manage nocturia in PD patients.

5.2   Causes for Nocturia in PD

The exact mechanism for the occurrence of nocturia in PD remains unclear, and several factors are found to be contributing to nocturia.

5.2.1   Reduced Bladder Capacity

Whenever the production of urine is excess of the bladder capacity, patients have to void. Therefore, patients with reduced bladder capacity at night are more likely to complain nocturia. Reduced bladder capacity in patients with PD can be caused by decreased compliance of the bladder muscle, involuntary contraction of the detrusor (detrusor overactivity, DO), and incomplete empty following a void [8]. DO is considered as an excess micturition reflex, and the micturition reflex is mediated by dopamine (both inhibitory in D1 receptors and facilitatory in D2 receptors) and GABA (inhibitory) pathways [9]. Nocturia in PD may be attributable to dysregulation of the dopamine D1-GABAergic direct pathway, which ultimately leads to failed inhibition of the micturition reflex and results in overactive bladder (OAB).

5.2.2   Nocturnal Polyuria

Nocturnal polyuria (NP) can be also associated with nocturia, which is characterized by excessive production of nocturnal urine. By definition, NP refers to that nocturnal urine production exceeds 20% of the entire 24-hour volume in young adults and 33% in the elderly [3]. Nocturnal polyuria are often resulting from systematic causes, such as dysregulated arginine vasopressin secretion, heart failure, kidney dysfunction, nocturnal fluid overload (especially alcohol and caffeine), use of long-acting diuretics, and sleep apnea [8].

5.2.3   Circadian Rhythm Disturbances

The circadian rhythm is regulated by the suprachiasmatic nucleus (SCN) of the hypothalamus, which controls the release of melatonin from the pineal gland in response to the environmental light/dark cycle [10]. However, structures involved in regulating circadian rhythm can be affected during the process of neurodegeneration and cell death in PD, including hypothalamus [10]. Moreover, in PD mouse model, overexpression of α-synuclein can cause a decreased SCN firing rate and potentially reduced their ability to transfer neural and hormonal signals from the central clock, which may also lead to circadian dysregulation [11]. Meanwhile, renal function and urine production are also modulated by circadian regulation of sodium and free water handling [12]. Furthermore, bladder capacity has also been found to be regulated by hormones regulating circadian cycle [13]. Therefore, circadian dysregulation in PD might lead to not only nocturnal polyuria but also decreased bladder capacity, which ultimately results in nocturia in PD.

5.2.4 Other Possible Mechanism for Nocturia in PD

Living habits such as excessive caffeine and alcohol intake, especially at night, can lead to nocturia. Other comorbidities such as renal dysfunction and diabetes can also result in nocturia.

5.3 Diagnosis and Evaluation of Nocturia in PD [14]

5.3.1 History Taking and Physical Examination

History taking is the first and most important step in nocturia assessment and should be enquired at every visit. Information on urine frequency (especially nocturnal), urinary urgency, incontinence, other comorbidities, and medication history should be gathered. It is important to distinguish nocturnal voiding is caused by bladder symptoms or because of other reasons, such as psychiatric problems. Moreover, asking patients to record their urine frequency and volume after each void as urinary diary is encouraged, which provides subjective and comprehensive information on patients' LUT symptoms and is the only effective method to figure out nocturnal polyuria. In this condition, measuring tools are needed to ensure the accuracy. Questionnaire is also desirable in assessing nocturia, such as PD sleep scale. Furthermore, doctors should also pay attention to the relationship between L-dopa use and LUT and motor fluctuations. Other medication uses such as long-acting diuretics can also contribute to nocturia. A review of other comorbidities would help to elucidate other system disorders which can cause nocturia, such as renal dysfunction and diabetes. Lifestyle factors should also be considered, such as caffeine and alcohol intake. Physical examination should include abdomen, pelvic, and genital organs, sacral cord mediated reflexes, anal sphincter tone and squeeze response.

5.3.2 Investigation

5.3.2.1 Routine Test
Routine blood test measuring serum creatinine and estimated glomerular filtration rate (eGFR) helps in evaluating kidney functions. Routine urine test is a useful way to exclude infections.

5.3.2.2 Ultrasonography
The post-void residual (PVR) urine volume can be measured by ultrasound or alternatively by in–out catheterization. Increase in PVR volume could indicate

voiding problem. However, ultrasonography is unable to differentiate whether incomplete voiding is resulting from difficulty in detrusor contract or urinary tract obstruction. In this condition, urodynamics is needed. Moreover, PVR volume is supposed to be recorded at different time to figure out the dynamic change of bladder emptying.

5.3.2.3 Urodynamics

Urodynamics is helpful in understanding the cause of LUT dysfunction in PD. It can help in figure out whether urine dysfunction is caused by reduced bladder compliance or detrusor overactivity. Notably, urodynamics should be performed ahead of any medication and can be applied to monitor treatment outcomes.

5.3.2.4 Other Investigations

Urethrocystoscopy (together with bladder washing cytology if possible) can be applied to detect urethrostenosis, urethral and bladder calculus, and bladder neoplasms.

5.4 Management and Treatment of Nocturia in PD

Nocturia in PD patients has detrimental influence on patients' quality of life; therefore, multiple disciplinary managements are needed to ameliorate symptoms. These include pharmacological management, lifestyle change, and urology referral when necessary.

5.4.1 Pharmacological Interventions

5.4.1.1 Dopamine Agonists

It is suggested that in patients with PD, the dopamine D1-GABAergic direct pathway which normally suppresses the micturition reflex is disrupted [9]. Therefore, dopamine agonists may have potential effects on nocturia in PD. Pergolide has been studied to improve nocturia [15], but it is restricted because of its side effects. Some open-label randomized control trials have indicated that rotigotine transdermal patch could provide beneficial effects to PD patients with nocturia [16].

5.4.1.2 Antimuscarinic Agents

Antimuscarinic agents are competitive antagonists for muscarinic acetylcholine receptors, which can alleviate overactivity of bladder and improve storage symptoms. Oxybutynin, tolterodine, solifenacin, darifenacin, fesoterodine, and trospium chloride are all belonging to this category [12]. However, when applying antimuscarinic agents, careful attention should to be paid to their side effects, including increasing anticholinergic burden such as urinary retention, constipation, and cognitive impairment.

5.4.1.3 Desmopressin

Desmopressin, an artificial synthesized form of 8-arginine vasopressin (ADH), is an antidiuretic drug which can temporarily suppress production of urine and overactivity of bladder detrusor. Therefore, it is useful for treating nocturnal polyuria. In a clinical trial that included eight patients complaining nocturia in PD, five patients showed clinically and statistically significant reduction in the frequency of nocturnal voids [17]. However, larger sample size and long-term follow-up studies are needed. Notably, hyponatremia or congestive heart failure should be carefully monitored in the patients prescribed with desmopressin. Also, patients older than 65 years old should be prescribed with caution.

5.4.1.4 Melatonin

Nocturia may be due to circadian rhythm disturbances of melatonin secretion; therefore, melatonin is a promising treatment for nocturia [18]. However, melatonin for nocturia in PD patients is poorly studied, and a phase 2 clinical of sustained-release melatonin 2 mg once daily for 12 weeks in patients with PD reporting nocturia is undergoing (clinicaltrials.gov/ct2/show/NCT02359448), which aims to evaluate the effects of exogenous melatonin on bother related to nocturia.

5.4.2 Intermittent Catheterization

Intermittent catheterization has been shown to be the major means for treating disorders which are related to incomplete bladder voiding. And because most of PD patients have bradykinesia as the dominant symptom, caregivers should be educated well for in-home catheterization.

5.4.3 Botulinum Toxin Injection

Previous studies have found intravesical botulinum toxin injection to be a safe and efficient in treating neurological DO; therefore, it is also a promising and helpful way for treating intractable DO in patients with PD. In a previous study which only enrolled four patients with PD, patients were administered with 200 U botulinum toxin type A into the bladder detrusor muscle at 20 different sites under the guidance of cystoscope and were reviewed 1 and 3 months after injection. Results showed significant decrease in daytime ($p < 0.0001$) and nighttime ($p < 0.0006$) urinary frequency in all four patients [19]. In another 6-month follow-up study, eight patients with PD were treated with intradetrusor injection of 100 U botulinum toxin type A. At the revisit at 1, 3, and 6 months after treatment, patients also showed significantly decrease in nighttime urinary frequency [20]. However, doctors should pay special attention to the risk for urinary retention, because of the potential difficulty to perform self-intermittent catheterization for patients with PD.

5.4.4 Surgical Treatment

Deep brain stimulation (DBS) of the globus pallidus pars interna (GPi) or the sub-thalamic nucleus (STN) have been established tools for managing motor symptoms in advanced PD patients. But its effect on nocturia remains inconsistent. In a previous study enrolled 107 advanced PD patients, when compared with those who did not underwent DBS, patients underwent DBS had a significant decreased nocturia ($p = 0.007$) [21]. However, in another study investigated whether DBS can improve LUT after a 12-month follow-up, results showed that nocturia did not improve significantly after either GPi DBS or STN DBS [22]. Therefore, more studies are needed to clarify DBS's effect on nocturia.

5.4.5 Neuromodulation

Transcutaneous tibial nerve stimulation (TTNS), which stimuli the tibial nerve by inserting a gauge needle, has been shown to be an efficient and safe treatment for idiopathic overactive bladder syndrome [23]. In a study enrolled 13 PD patients, patients treated with TTNS showed significant reduction in nocturia episodes (4.0 vs 2.0; $p < 0.01$) after 10 treatments [24]. Studies specifically evaluating the outcomes of sacral neuromodulation in PD are lacking. In a retrospective study, it was found that 4 of 6(67%) patients with PD were found to have decrease of nocturia of 70% [25].

5.4.6 Lifestyle Change

It is important to restrict fluid intake in the treatment of nocturia. First, reducing water, coffee, alcohol, and other fluid intake a few hour before bedtime is helpful [26]. Also, empty the bladder before sleep is also necessary. Besides, it is also helpful to apply diuretics in the late afternoon [27]. Furthermore, nocturia may also attributable to absorption of dependent edema fluid at night; therefore, exercise, elevating lower limb above the heart level in the afternoon, and use of compression stockings are also encouraged in PD patients with nocturia [28].

References

1. Blackett H, Walker R, Wood B. Urinary dysfunction in Parkinson's disease : a review. Parkinsonism Relat Disord. 2009;15(2):81–7.
2. Madan A, Ray S, Burdick D, Agarwal P. Management of lower urinary tract symptoms in Parkinson's disease in the neurology clinic. Int J Neurosci. 2017;127(12):1136–49.
3. Van KP, Abrams P, Chaikin D, Donovan J, Fonda D, Jackson S, et al. The standardisation of terminology in nocturia : report from the Standardisation Sub-committee of the International Continence Society. BMJ Int. 2002;183(2):179–83.

4. Rana AQ, Vaid H, Akhter MR, Awan NY, Fattah A, Cader MH, et al. Prevalence of nocturia in Parkinson's disease patients from various ethnicities. Neurol Res. 2014;36(3):234–8.
5. Rana AQ, Paul DA, Qureshi AM, Ghazi A, Alenezi S, Rana MA, et al. Association between nocturia and anxiety in Parkinson's disease. Neurol Res. 2015;37(7):563–7.
6. Yeo L, Singh R, Gundeti M. Urinary tract dysfunction in Parkinson's disease : a review. Int Urol Nephrol. 2012;44(2):415–24.
7. Sakakibara R, Panicker J, Finazzi-agro E, Iacovelli V, Bruschini H. A guideline for the management of bladder dysfunction in Parkinson's disease and other gait disorders. Neurourol Urodyn. 2016;35(5):551–63.
8. Batla A, Phe V, De ML, Panicker JN. Nocturia in Parkinson's disease : why does it occur and how to manage ? Mov Disord Clin Pract. 2016;3(5):443–51.
9. Sakakibara R, Tateno F, Nagao T, Yamamoto T, Uchiyama T, Yamanishi T, et al. Bladder function of patients with Parkinson's disease. Int J Urol. 2014;21(7):638–46.
10. Easton A, Ph D, Meerlo P, Ph D, Bergmann B, Ph D, et al. The suprachiasmatic nucleus regulates sleep timing and amount in mice. Sleep. 2004;27(7):1307–18.
11. Kudo T, Loh DH, Truong D, Wu Y, Colwell CS. Circadian dysfunction in a mouse model of Parkinson's disease. Exp Neurol. 2011;232(1):66–75.
12. Buser N, Ivic S, Kessler TM, Kessels AGH, Bachmann LM. Efficacy and adverse events of antimuscarinics for treating overactive bladder : network meta-analyses. Eur Urol. 2012;62(6):1040–60.
13. Matsuta Y, Yusup A, Tanase K, Ishida H, Akino H. Melatonin increases bladder capacity via GABAergic system and decreases urine volume in rats. J Urol. 2010;184(1):386–91.
14. Panicker JN, Fowler CJ, Kessler TM. Lower urinary tract dysfunction in the neurological patient : clinical assessment and management. Lancet Neurol. 2015;14(7):720–32.
15. Kuno S, Mizuta E, Yamasaki S, Araki I. Effects of pergolide on nocturia in Parkinson's disease: three female cases selected from over 400 patients. Parkinsonism Relat Disord. 2004;10(3):181–7.
16. Rosa-grilo M, Qamar MA, Chaudhuri KR. Rotigotine transdermal patch and sleep in Parkinson's disease : where are we now? NPJ Parkinson's Dis. 2017;3:28.
17. Suchowersky O, Furtado S, Rohs G. Beneficial effect of intranasal desmopressin for nocturnal polyuria in Parkinson's disease. Mov Disord. 1995;10(3):337–40.
18. Marinkovic SP, Gillen LM, Stanton SL. Managing nocturia. BMJ. 2004;328(7447):1063–6.
19. Giannantoni A, Rossi A, Mearini E, Del ZM, Porena M, Berardelli A. Botulinum toxin A for overactive bladder and detrusor muscle overactivity in patients with Parkinson's disease and multiple system atrophy. J Urol. 2009;182(4):1453–7.
20. Giannantoni A, Conte A, Proietti S, Giovannozzi S, Rossi A, Fabbrini G, et al. Botulinum toxin type A in patients with Parkinson's disease and refractory overactive bladder. J Urol. 2011;186(3):960–4.
21. Winge K, Nielsen KK. Bladder dysfunction in advanced Parkinson's disease. Neurourol Urodyn. 2012;31(8):1279–83.
22. Witte LP, Odekerken VJJ, Boel JA, Schuurman PR, Gerbrandy-Schreuders LC, de Bie RMA, et al. Does deep brain stimulation improve lower urinary tract symptoms in Parkinson's disease? Neurourol Urodyn. 2018;37(1):354–9.
23. Booth J, Lawrence M. The effectiveness of transcutaneous tibial nerve stimulation (TTNS) for adults with overactive bladder syndrome : a systematic review. Neurourol Urodyn. 2018;37(2):528–41.
24. Carolina M, Carlos P, Levi A, Martins R, Anelyssa C, Lucio A. Transcutaneous tibial nerve stimulation in the treatment of lower urinary tract symptoms and its impact on health-related quality of life in patients with Parkinson disease: a randomized controlled trial. J Wound Ostomy Cont Nurs. 2015;42(1):94–9.
25. Wallace PA, Lane FL, Noblett KL. Sacral nerve neuromodulation in patients with underlying neurologic disease. Am J Obstet Gynecol. 2007;197(1):96.e1–5.

26. Griffiths DJ, Mccracken PN, Harrison GM, Ann Gormley E. Relationship of fluid intake to voluntary micturition and urinary incontinence in geriatric patients. Neurourol Urodyn. 1993;12(1):1–7.
27. Weiss JP, Blaivas JG. Nocturia. J Urol. 2000;163(1):5–12.
28. Batla A, Tayim N, Pakzad M, Panicker JN. Treatment options for urogenital dysfunction in Parkinson's disease. Curr Treat Options Neurol. 2016;18(10):45.

Restless Legs Syndrome (RLS) and Periodic Limb Movement Disorder

6

Kang-Ping Xiong and Chun-Feng Liu

Abstract

Restless legs syndrome (RLS) is a very common sleep-related movement disorder which was characterized by an urge to move the legs frequently accompanied by uncomfortable and unpleasant sensations that are difficult to describe. Periodic limb movements during sleep (PLMS) are repetitive, highly stereotypic limb movements. Clinically, PD and RLS share several features, such as responsiveness to dopaminergic pharmacotherapy, exacerbations by dopaminergic antagonists, and association with PLMS. The prevalence of RLS in PD patients is very variable which is ranging approximately from 0 to 50%. It is still unclear if PD and RLS are etiologically related. RLS might be an early clinical feature of incident PD. There are validated genetic risk factors that are associated with both RLS and PD. The dopaminergic dysfunction is crucial in pathophysiology of both conditions. Diagnosis of RLS in PD patients is challenged because of the potential overlap that included evening rigidity and dyskinesia, early morning dystonia, and sensory disturbances. Until now, there are no specific diagnostic criteria for RLS in PD patients. The aim of treatment for RLS in PD is applying safety and effectiveness including both pharmacologic and nonpharmacologic therapies.

Keywords

Restless legs syndrome · Periodic limb movement disorder · Genetic markers · Pathophysiology · Imaging · Diagnosis · Treatment

Electronic Supplementary Material The online version of this chapter (https://doi.org/10.1007/978-981-15-2481-3_6) contains supplementary material, which is available to authorized users.

K.-P. Xiong · C.-F. Liu (✉)
Department of Neurology, The Second Affiliated Hospital of Soochow University, Suzhou, China
e-mail: liuchunfeng@suda.edu.cn

Case

The patient, a 72-year-old retired worker, was diagnosed with Parkinson's disease for 3 years ago and started with anti-parkinsonism medication after diagnosis. Two years later, he complained uncomfortable and unpleasant sensations such as numbness, tumidness, pain on his legs which made him very urge to move his legs, and symptoms existed very frequently during night and daily nap and even while taking rest, and the abnormal sensations were triggered by lying down or sitting down for a while, resulting in difficulty of falling asleep. These kinds of abnormal sensations would be partially or totally relieved by moving or stretching legs or walking out of bed, at least as long as the movements continued. The doctor diagnosed him with RLS and helped him to perform a video-PSG. His symptoms completely and quickly resolved with the prescription of a low-dose dopamine agonist 1 h before bedtime and he had no motor fluctuation or dystonia, suggesting that he suffered typical RLS. The periodic limb movement during wake (PLMW) (Fig. 6.1) and PLMS (Fig. 6.2) from PSG of this patient had been attached as follows.

6.1   Introduction

Restless legs syndrome (RLS) is a very common sleep-related movement disorder which was characterized by an urge to move the legs frequently accompanied by uncomfortable and unpleasant sensations that are difficult to describe. Patients complain their symptoms as numb, burning, twitching, or pain in their legs. However, in the most severe cases the upper limbs can also be involved [1, 2]. Onset of symptoms occur frequently during rest or inactivity, and an exacerbation of uncomfortable sensations would happen when patients go to bed for sleep. This clear circadian trend with a peak in the evening or at night of RLS symptom can obviously result in poor nocturnal sleep [2]. A recent study demonstrated that prevalent RLS is related with higher risk of incident PD during 8 years of follow-up from a nationally representative prospective cohort of almost 3.5 million US veterans, suggesting that RLS could be an early clinical feature of PD [3].

The prevalence of RLS arranged from 3.9 to 14.3%, and women are more affected than men and an increase with age by using minimal diagnostic criteria of the international restless legs syndrome study group (IRLSSG) [2, 4]. The prevalence of RLS seems to be different linked to geographic areas: highest in European populations (5–12%), intermediate in Asian countries (1–8%), and lowest in African countries (<1%) [5]. There is a variable prevalence of RLS in PD patients ranging approximately from 0 to 50% according to cross-sectional studies [2]. These variances of the prevalence were from the use of dopaminergic medicines in PD patients which may lead to an underestimation of RLS. On the other hand, the anti-parkinsonism medicines, especially Levodopa, may cause an increased frequency of "mimics" situations of RLS in PD [2].

Periodic leg or limb movements during sleep (PLMS) were first observed in the 1960s and the criteria for PLMS were generated in the early 1980s. PLMS are highly repetitive and stereotyped limb movements that occur during sleep, typically

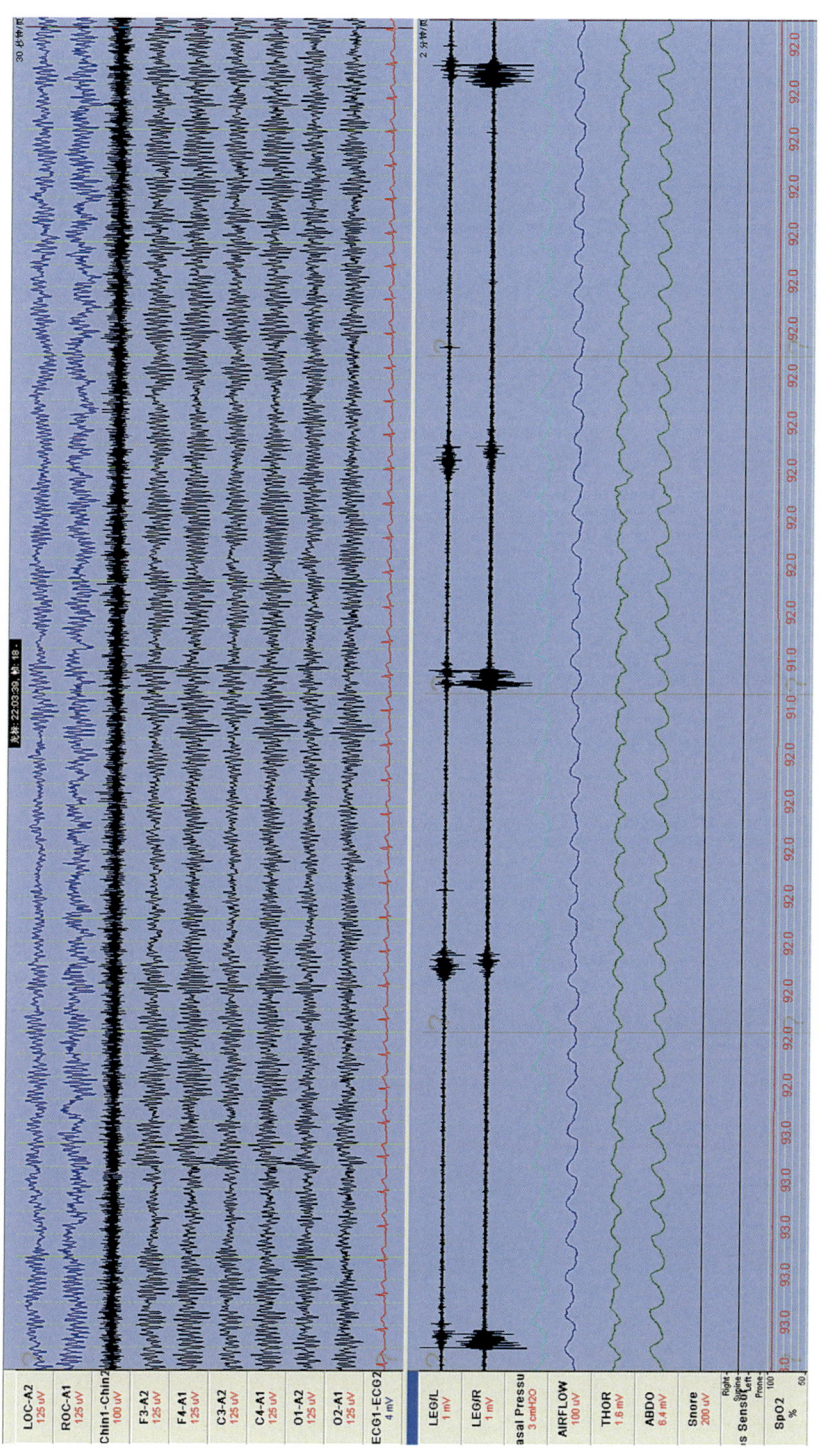

Fig. 6.1 PLMW

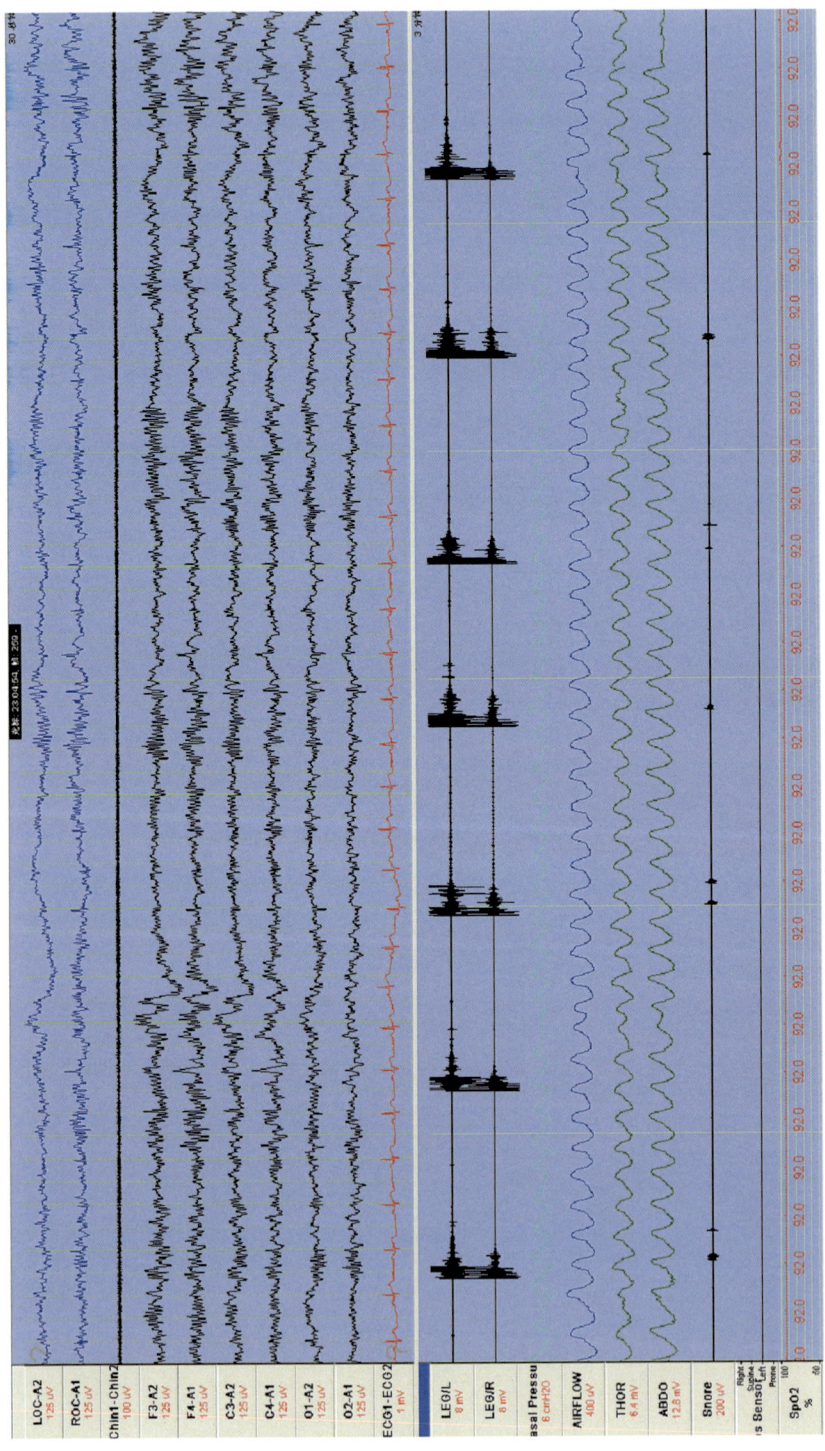

Fig. 6.2 PLMS

including the dorsal extension of the big toe, often accompanied with partial flexion of the ankle, the knee, and even the hip [6]. Periodic limb movement disorder (PLMD) is listed among the category of sleep-related movement disorders in the third revision of the International Classification of Sleep Disorders (ICSD-3) [1]. Previous studies have demonstrated that periodic leg movements index (PLMI) in treated PD patients is higher compared with controls, but there is no difference between de novo diagnosed PD patients and controls [6–8] and PLMI increases with advancing PD disease situation [9].

6.2   The Association Between PD and RLS

PD and RLS share several features, such as responsiveness to dopaminergic pharmacotherapy, exacerbations by dopaminergic antagonists, and association with PLMS. It is still unclear if PD and RLS are etiologically related. There are kinds of debatable issues including if RLS may precede the onset of PD or if RLS could represent a secondary condition of PD and if impaired dopaminergic pathway may represent a bridge between PD and RLS. The main pathophysiological hypotheses of the relationship between PD and RLS are as follows [2]: (1) two diseases may share the same pathophysiological mechanism including dopaminergic bridge, iron hypothesis, and non-dopaminergic hypothesis. (2) RLS in PD has a different pathophysiology from idiopathic RLS. There were sensory fluctuations in RLS symptoms related to Parkinsonism factors. (3) These two diseases are different entities.

RLS and PD respond positively to dopaminergic therapy suggested that the dopaminergic system may play a crucial role in both disorders. The occurrence of RLS in PD might be a consequence of dopaminergic therapies and RLS-like symptoms may be part of sensory-motor spectrum of wearing off [2]. PD may be a risk factor to develop RLS in combination with low ferritin levels [2]. A worse nutritional condition may lead to an iron deficiency in PD patients who exhibited RLS. The time of RLS occurrence and gender differences in PD patients with RLS are uncertain [2]. Men with RLS are more likely to have concurrent PD [10]. PD patients with RLS less likely to have a family history of RLS and the RLS symptoms seem to be less severe than idiopathic RLS [2, 11].

6.3   Genetic Markers of RLS in PD

There are well-validated genetic risk factors that are associated with each disorder for RLS and PD. Shared genetic loci may suggest shared pathophysiology; however, it is still unclear if these two disorders are etiologically related [12]. In the RLS GWAS, pleiotropy with PD was examined, and it was demonstrated that there is no pleiotropy between RLS and PD [13–15]. It has been examined the four well-validated RLS genetic risk markers in the genes MEIS1, BTBD9, PTPRD, and MAP 2K5/SKOR1 have no clear function in PD susceptibility [12]. Single-nucleotide polymorphism (SNPs) variants like TOX3 variants with RLS and PD are likely involved in the sex differences [15].

6.4   Pathophysiology of RLS in PD

The dopaminergic dysfunction is crucial in pathophysiology of both conditions. The pathophysiology of RLS has been implicated central dopaminergic dysfunction in several regions including the substantia nigra, striatum, putamen, and downstream disinhibition of the sensory dorsal horn and the intermediolateral nucleus of the spinal cord [16]. A study of investigating dopamine transporter using single photon emission computed tomography (FP-CIT SPECT) demonstrated that more dopamine transporters were preserved at the head of caudate in PD patients with RLS, indicating that there might be a preserved nigrostriatal dopaminergic pathway in comparison to patients without RLS [2, 17]. Postmortem studies of idiopathic RLS patients founded the absence of PD pathological biomarker, a-synuclein accumulation and Lewy bodies. In a large family that presented with Parkinsonism, essential tremor, RLS, and depression, only sparse Lewy bodies were found [12, 18]. Recently, the specific dysfunction of iron transportation in the brain and the iron–dopamine interaction have been studied [6]. There is not only elevated glutamatergic activity in the thalamus but also adenosine A1 receptor downregulation plays a role in the hyperarousal of RLS and in PLMS [6]. This downregulation is secondary to iron deficiency. The role of central hypoxic pathways and peripheral tissue hypoxia in RLS is another object of current research [6].

6.5   Imaging of RLS in PD

Brain imaging studies investigating dopaminergic dysfunction in RLS patients have not yielded conclusive results. SPECT/PET studies have shown that nigrostriatal functions and ligand binding to the striatal dopamine transporters (DAT) and D2 receptors in RLS are normal, while other studies have found a reduced ability of D2 receptors to bind ligands and decreased 18F-dopa uptake in the striatum and putamen [16]. The iron content in the red nuclei and the substantia nigra was reduced in the RLS patients [16]. Iron in the substantia nigra is increased of PD and reduced in RLS [12, 16, 19]. Decreased echogenicity in the substantia nigra has been reported in RLS patients compared with PD patients and healthy controls [15]. Compared with idiopathic RLS and controls, a significantly increased area of echogenicity in the substantia nigra was found in subjects with both PD and RLS, demonstrating the existence of different mechanisms for regulating brain iron in idiopathic RLS and PD patients with RLS [15]. No significant difference in substantia nigra echogenicity was found between PD patients with and without RLS [1, 20, 21].

6.6   Diagnosis of RLS in PD

The third revision of the International Classification of Sleep Disorders (ICSD-3) defines RLS as an "urge to move the legs, usually accompanied by uncomfortable or unpleasant sensations in the legs." These symptoms must (1) begin or worsen during periods of rest in activities such as lying down or sitting; (2) be partially or totally relieved by movement, such as walking or stretching, at least as long as the activity continues; and (3) occur exclusively or predominantly in the evening or night rather than during the day. The ICSD-3 also requires that these features cannot be explained by another medical or behavioral condition, such as leg edema, arthritis, leg cramps, positional discomfort, myalgia, venous stasis, or habitual foot tapping [1, 6].

It is difficult for diagnosing RLS in PD patients because of the potential overlap with evening rigidity, dyskinesia, early morning dystonia, and PD sensory symptoms [6]. Until now, there are no specific diagnostic criteria for RLS in PD patients. It has been suggested that the suggested immobilization test (SIT) might be useful which is an assessment of leg discomfort and PLMS during a period of wakening in the evening with the legs held still. The SIT was able to discriminate PD patients with RLS from PD patients without RLS (whereas a PLMI was not able to make this distinction), with a mean leg discomfort cutoff of 11 showing sensitivity of 91% and specificity of 72% for RLS diagnosis in PD [6, 22].

Leg movement (LM) events are defined as a duration of 0.5–10 s [6]. PLM are defined as at least 4 LMs, with a period length of 5–90 s between two LMs [6, 23]. The official World Association of Sleep Medicine (WASM) standards for recording and scoring periodic leg movements in sleep (PLMS) and wakefulness (PLMW) were published in 2006 for objective evaluation [24, 25]. Diagnostic criteria for PLMD include PLMS in the PSG, with a frequency >5/h in children and >15/h in adults. These criteria also require that the PLMS cause clinically sleep disturbance or function impairment and are not better explained by another current medical or neurological disorder [6]. Thus, the ICSD-3 states that PLMD cannot be diagnosed in the context of RLS, narcolepsy, untreated obstructive sleep apnea, or RBD. The diagnosis of RLS takes precedence over that of PLMD when potentially sleep-disrupting PLMS occur in the context of RLS [1, 6].

6.7   Treatment of RLS in PD

RLS is typically a chronic disease and requires a long period treatment. The treatment aim of RLS is applying safe and effective therapies including both pharmacologic and nonpharmacologic approaches, to relieve RLS symptoms and improve quality of life [26]. Lifestyle adjustment is recommended for PD patients with mild RLS. Other nonpharmacological treatments include massaging the affected limbs, bathing in hot or cold water, physical activity, and distraction therapy.

Concomitant medications that may induce or aggravate RLS symptoms (e.g., antidopaminergic drugs, antihistamines, and antidepressants) should be stopped whenever possible. In PD patients with RLS, other secondary factors and contributing comorbidities should be excluded such as metabolic disorders, end-stage renal disease, diabetes, pregnancy, and serotonergic antidepressants. Treating the iron deficiency should be the first line of treatment. If the serum level of ferritin is <50–75 µg/ml or transferrin saturation is <20%, oral iron supplementation is recommended. If oral iron is not tolerated or is contraindicated, intravenous iron could be considered.

Different drugs have shown a high efficacy. In particular, dopamine agonists are effective in relieving patients' symptoms and are considered first-line treatment [27]. To prevent augmentation, the lowest possible effective dose of dopaminergic agents is recommended and long-acting DAs are preferred. Rotigotine patch is strongly recommended in the management of RLS in PD Patients [28]. Alpha-2-delta agonists (gabapentin, enacarbil, and pregabalin) are recognized as a valid alternative [29]. For patients with PD and moderate to severe RLS, subthalamic nucleus (STN) deep brain stimulation (DBS) improves RLS symptoms [30].

References

1. American Academy of Sleep Medicine. The international classification of sleep disorders: diagnostic and coding manual. 3rd ed. rev. ed. Darien, IL: American Academy of Sleep Medicine; 2014.
2. Ferini-Strambi L, Carli G, Casoni F, Galbiati A. Restless legs syndrome and Parkinson disease: a causal relationship between the two disorders? Front Neurol. 2018;9(6):551. https://doi.org/10.3389/fneur.2018.00551.
3. Szatmari S, Bereczki D, Fornadi K, et al. Association of restless legs syndrome with incident Parkinson's disease. Sleep. 2017;40(2):zsw065.
4. Ohayon MM, O'hara R, Vitiello MV. Epidemiology of restless legs syndrome: a synthesis of the literature. Sleep Med Rev. 2012;16:283–95. https://doi.org/10.1016/j.smrv.2011.05.002.
5. Koo BB. Restless leg syndrome across the globe: epidemiology of the restless legs syndrome/Willis-Ekbom disease. Sleep Med Clin. 2015;10:189–205. https://doi.org/10.1016/j.jsmc.2015.05.004.
6. Högl B, Stefani A. Restless legs syndrome and periodic leg movements in patients with movement disorders: specific considerations. Mov Disord. 2017;32(5):669–81.
7. Wetter TC, Collado-Seidel V, Pollmacher T, Yassouridis A, Trenkwalder C. Sleep and periodic leg movement patterns in drug-free patients with Parkinson's disease and multiple system atrophy. Sleep. 2000;23:361–7.
8. Wetter TC, Brunner H, Högl B, Yassouridis A, Trenkwalder C, Friess E. Increased alpha activity in REM sleep in de novo patients with Parkinson's disease. Mov Disord. 2001;16:928–33.
9. Young A, Home M, Churchward T, Freezer N, Holmes P, Ho M. Comparison of sleep disturbance in mild versus severe Parkinson's disease. Sleep. 2002;25:573–7.
10. Gao X, Schwarzschild MA, O'Reilly EJ, Wang H, Ascherio A. Restless legs syndrome and Parkinson's disease in men. Mov Disord. 2010;25:2654–7. https://doi.org/10.1002/mds.23256.
11. Bhalsing K, Suresh K, Muthane UB, Pal PK. Prevalence and profile of restless legs syndrome in Parkinson's disease and other neurodegenerative disorders: a case-control study. Parkinsonism Relat Disord. 2013;19:426–30. https://doi.org/10.1016/j.parkreldis.2012.12.005.
12. Gan-Or Z, Alcalay RN, Bar-Shira A, Leblond CS, Postuma RB, Ben- Shachar S, Waters C, Johnson A, Levy O, Mirelman A, Gana-Weisz M, Dupré N, Montplaisir J, Giladi N, Fahn S, Xiong L, Dion PA, Orr-Urtreger A, Rouleau GA. Genetic markers of restless legs syndrome in Parkinson disease. Parkinsonism Relat Disord. 2015;21(6):582–5. https://doi.org/10.1016/j.parkreldis.2015.03.010.
13. Schormair B, Zhao C, Bell S, et al. Identification of novel risk loci for restless legs syndrome in genome-wide association studies in individuals of European ancestry: a meta-analysis. Lancet Neurol. 2017;16(11):898–907. https://doi.org/10.1016/S1474-4422(17)30327-7.
14. Chang D, Nalls MA, Hallgrimsdottir IB, et al. A meta-analysis of genome-wide association studies identifies 17 new Parkinson's dis- ease risk loci. Nat Genet. 2017;49(10):1511–6. https://doi.org/10.1038/ng.3955.

15. Mohtashami S, He Q, Ruskey JA, et al. TOX3 variants are involved in restless legs syndrome and Parkinson's disease with opposite effects. J Mol Neurosci. 2018;64:341. https://doi.org/10.1007/s12031-018-1031-4.
16. Suzuki K, Miyamoto M, Miyamoto T, et al. Restless legs syndrome and leg motor restlessness in Parkinson's disease. Parkinsons Dis. 2015;2015:490938. https://doi.org/10.1155/2015/490938.
17. Moccia M, Erro R, Picillo M, Santangelo G, Spina E, Allocca R, et al. A four-year longitudinal study on restless legs syndrome in Parkinson disease. Sleep. 2016;39:405–12. https://doi.org/10.5665/sleep.5452.
18. Puschmann A, Pfeiffer RF, Stoessl AJ, Kuriakose R, Lash JL, Searcy JA, et al. A family with Parkinsonism, essential tremor, restless legs syndrome, and depression. Neurology. 2011;76:1623e30.
19. Garcia-Borreguero D, Odin P, Serrano C. Restless legs syndrome and PD: a review of the evidence for a possible association. Neurology. 2003;61:S49e55.
20. Kwon D-Y, Seo W-K, Yoon H-K, et al. Transcranial brain sonography in Parkinson's disease with restless legs syndrome. Mov Disord. 2010;25(10):1373–8.
21. Ryu JH, Lee MS, Baik JS. Sonographic abnormalities in idiopathic restless legs syndrome (RLS) and RLS in Parkinson's disease. Parkinsonism Relat Disord. 2011;17(3):201–3.
22. De Cock VC, Bayard S, Yu H, et al. Suggested immobilization test for diagnosis of restless legs syndrome in Parkinson's disease. Mov Disord. 2012;27:743–9.
23. The Atlas Task Force. Recording and scoring leg movements. Sleep. 1993;16:748–59.
24. Zucconi M, Ferri R, Allen R, et al. The official World Association of Sleep Medicine (WASM) standards for recording and scoring periodic leg movements in sleep (PLMS) and wakefulness (PLMW) developed in collaboration with a task force from the International Restless Legs Syndrome Study Group (IRLSSG). Sleep Med. 2006;7:175–83.
25. Michaud M, Lavigne G, Desautels A, Poirier G, Montplaisir J. Effects of immobility on sensory and motor symptoms of restless legs syndrome. Mov Disord. 2002;17:112–5.
26. Liu CF, Wang T, Zhan SQ, et al. Management recommendations on sleep disturbance of patients with Parkinson's disease. Chin Med J. 2018;131(24):2976–85.
27. Loddo G, Calandra-Buonaura G, Sambati L, Giannini G, Cecere A, Cortelli P, et al. The treatment of sleep disorders in Parkinson's disease: from research to clinical practice. Front Neurol. 2017;8:42. https://doi.org/10.3389/fneur.2017.00042.
28. Wang Y, Yang YC, Lan DM, Wu H, Zhao ZX. An observational clinical and video-polysomnographic study of the effects of rotigotine in sleep disorder in Parkinson's disease. Sleep Breath. 2017;21:319–25. https://doi.org/10.1007/s11325-016-1414-0.
29. Iftikhar IH, Alghothani L, Trotti LM. Gabapentin enacarbil, pregabalin and rotigotine are equally effective in restless legs syndrome: a comparative meta-analysis. Eur J Neurol. 2017;24:1446–56. https://doi.org/10.1111/ene.13449.
30. Klepitskaya O, Liu Y, Sharma S, et al. Deep brain stimulation improves restless legs syndrome in patients with Parkinson disease. Neurology. 2018;00:e1–9. https://doi.org/10.1212/WNL.0000000000006162.

Sleep-Related Breathing Disorders

7

Xiao-jing Gu, Bei Cao, and Hui-fang Shang

Abstract

Sleep-related breathing disorder (SBD) is one of the several chronic disorders, which is characterized by snoring and partial or complete suspension of breathe during sleep. SBD can be classified into obstructive sleep apnea (OSA) and central sleep apnea (CSA). There are bidirectional relationship between SBD and Parkinaon's disease (PD), which means that SBD is a manifestation of PD, while SBD is also involved in the development and progression of PD. The mechanism of SBD in PD remain poorly understood. Similar to the general population, aging, male, and obesity have been found to increase the risk for developing SDB in patients with PD. Moreover, PD itself is contributed to the pathogenesis of SDB. In order to diagnose SBD in PD patients, detailed sleep history, medical history, life habits, and some sleep-related scales are needed; more importantly, polysonomograph is the golden standard for diagnosing SBD. Management of SBD in PD needs comprehensive methods, including change of life habits, medication, and continuous positive airway pressure therapy (CPAP).

Keywords

Sleep-related breathing disorder · Obstructive sleep apnea · Central sleep apnea · Polysonomograph · Medication · Continuous positive airway pressure therapy

Case

The patient, a 54-year-old worker who worked in a manufacturing factory, was diagnosed with Parkinson's disease (PD) 5 years ago and started with medication. His wife reported that the patient started to snore recently when the patient had

X.-j. Gu · B. Cao · H.-f. Shang (✉)
Department of Neurology, West China Hospital, Sichuan University,
Wuhou District, Chengdu, Sichuan, China

© Springer Nature Singapore Pte Ltd. 2020
C.-F. Liu (ed.), *Sleep Disorders in Parkinson's Disease*,
https://doi.org/10.1007/978-981-15-2481-3_7

overworked in the day. The doctor suggested the patient to sleep in the lateral position and raise the bed head. Three months later, the patient's wife complained that the patient could not sleep in the lateral position because his rigidity and postural instability worsened, so his snoring became more severe and more frequent at night. Moreover, the patient was taking clonazepam at night because he had rapid eye movement behavior disorder. Then his doctor advised him to stop taking clonazepam and prescribed the patient with controlled-release Sinemet 100 mg per night. The patient was reviewed again after 2 months and his snoring was relieved. Half a year later, the patient complained that he sometimes could not breath during sleep and was forced to wake up. His wife said that the patient had breathing movement during cessation of breathing. Also the patient complained that he had headache and felt thirsty in the morning after waking up. Moreover, he felt extremely sleepy in the daytime such as when watching TV and taking the bus, and he could not focus on his work. Polysonomograph was prescribed and the result showed the patient had an apnea/hypopnea index of 13, which suggested he had moderate obstructive sleep apnea (OSA). He was advised to the otorhinolaryngological department, and otolaryngological examination and pharyngoscopy screening showed he did not had structural abnormality in the upper airway. Then his OSA was diagnosed to be primary OSA in PD. He was suggested to take noninvasive positive pressure ventilation at night. At review after 6 months, he was doing well.

7.1   Introduction

Sleep-related breathing disorder (SBD), also known as sleep disordered breathing (SDB), is defined by several chronic conditions characterized by snoring and partial or complete cessation of breathing during sleep. SBD can be classified into obstructive sleep apnea (OSA) and central sleep apnea (CSA) [1]. OSA is often caused by the repetitive collapse of the upper airway during sleep. Patients with OSA are often manifested as a cessation or a significant reduction in airflow in the presence of breathing effort, while CSA is due to the dysregulation of the respiratory center in the brainstem, and patients with CSA is characterized by absence of airflow and thoracoabdominal breathing movement at the same time [1]. Pure CSA is very rare in patients with PD, while the prevalence of OSA varies from 20 to 60% among different studies [2]. This great heterogeneity in prevalence of OSA is likely to be caused by variable sample size of each study and also selection bias. Furthermore, non-uniform scoring system of respiratory events between different laboratories also contribute to the discrepancy. Regarding the prevalence of OSA in PD, on the one hand, some studies have found that compared to non-PD population, OSA is more prevalent in patients with PD [3–5], while other studies found no increase of developing OSA in PD patients [6, 7]. Studies have shown that SBD and PD have a bidirectional relationship, which suggests that SBD is a manifestation of PD, while SBD is also involved in the development and progression of PD [8]. Moreover, patients with OSA were more likely to develop cognition impairment, excessive daytime sleepiness, and psychiatric problems, which are deleterious on patients' and caregivers' quality of life [9]. Therefore, SBD in PD patients needs special attention.

7.2   Causes of SBD in PD

The mechanism of SBD in PD remain poorly understood. Similar to the general population, aging, male, and obesity have been shown to increase the risk for developing SDB in patients with PD. Moreover, PD itself is contributed to the pathogenesis of SDB. Hypokinesia and rigidity in PD can affect the upper airway musculature, which results in upper airway obstruction. These conditions could get deteriorated during sleep, which finally leads to OSA [8]. Furthermore, autonomic dysfunction can also occur in patients with PD, which may lead to the impairment of breathing controlling. And this phenomenon can become particularly obvious during non-REM sleep period, during when respiration is mainly depending on chemical drive [8]. Furthermore, sleep fragmentation itself can lead to the disturbance of respiration. In humans, sleep fragmentation can induce upper airway collapse during sleep, which results in SDB in PD patients [8]. Moreover, although it is well recognized that loss of dopaminergic neurons in substantia nigra is the hallmark neurodegeneration process happening in PD, neurodegeneration itself can also affect other brain regions, such as the hypothalamus and brainstem [10]. The hypothalamus plays an especially important role in the process of sleep, with many sleep-promoting nuclei located in this region and also many neurochemical efferent fibers emanating from this region, for example, sleep-active neurons of the ventrolateral and median preoptic areas. Therefore, neurodegenerative process of sleep-related central nervous system can lead to CSA in PD patients [2].

7.3   Diagnosis [11]

7.3.1 History Taking and Physical Examination

It is of great importance and necessity to take a detailed sleep history in PD patients. Patients, and more importantly, bed partners, family members, and caregivers should be asked. Questions include frequency and severity of snoring and cessation of apnea during sleep, witnessed apnea, gasping/choking episodes, total sleep amount, relieving and predisposing factors, morning fatigue or headache, and excessive daytime sleepiness. A sleep diary can guide patients to subjectively record down their sleep conditions. It should include daily sleep and awake hours and should be consecutively kept for more than 2 weeks. The sleep diary can provide clinicians with insights into the patients' total amount of sleep in a 24-h period, and also it allows doctors to understand the patient's sleep pattern. Also, medication history is needed, because medications such as sedatives and neuroleptics can lead to SDB. Routine and systematic physical examination is also needed, which include weight, height, body mass index (BMI), neck circumference, blood pressure, and heart rate; a thorough nasal and upper airway examinations; and auscultation for any cardiac or respiratory abnormalities. In patients with OSA, upper airway stenosis and occlusion can be found on examination. Also, it is of great importance to observe if there is any absence of breathing efforts during apnea. Moreover, questionnaires can be applied when possible, such as Sleep Apnea Scale of the Sleep Disorders Questionnaire (SA-SDQ), Multivariable Apnea Prediction (MAP) Index, Berlin Questionnaire, STOP-Bang Questionnaire, Calgary Sleep Apnea Quality of Life Index (SAQLI), and Quebec Sleep Questionnaire (QSQ) [12].

7.3.2 Investigations

The gold standard of diagnosing SDB should be based on polysonomograph. According to the American Academy of Sleep Medicine (AASM) Manual for the Scoring of Sleep and Associated Events [13], SBD was diagnosed based on apnea-hypopnea index (AHI) > 5/h, which was calculated by the average number of apneas and hypopneas occurring per hour. By definition, apnea refers to a decrease in airflow of ≥90% (compared with baseline) lasting more than 10 s, while hypopnea refers to a decrease in airflow of ≥30% (by a valid measure of airflow) for more than 10 s, associated with either ≥3% desaturation from the pre-event baseline or an arousal.

OSA mainly results from the repetitive collapse of the upper airway during sleep and is characterized by a cessation or a significant reduction in airflow, but with thoracoabdominal breathing movement. If OSA accounts for more than 50% of all apneic events, then it's classified as OSA.

CSA is due to the dysregulation of the respiratory center in the brainstem, and patients with CSA is characterized by absence of airflow and thoracoabdominal breathing movement at the same time. The PSG diagnosis criteria for CSA is that CSA accounts for more than 50% of all apneic events.

The severity of SBD can be graded as mild (AHI: 5–15/h), moderate (AHI: 15–30/h), and severe (AHI > 30/h).

7.4 Treartment [14, 15]

The management strategies for SDB in PD patients are almost the same as treating SDB in the non-PD population, which should be treated as a chronic condition that requires long-term and multidisciplinary management. The therapeutic goal is to decrease the frequency and severity of SBD, alleviate SBD symptoms, and improve SBD-induced cognition impairment, excessive daytime sleep, and psychiatric problems [16].

7.4.1 Lifestyle Change

Lifestyle change should include weight loss (ideally reduce BMI to at least 25 kg/ m^2); physical exercise; avoidance of alcohol and smoking; cautious with the sedatives before bedtime (e.g., when treating PD patients with SDB and RBD, clonazepam should be avoided); sleep in the non-supine position; raising the bed head; and avoidance of overwork in the daytime.

7.4.2 Medication

Dopaminergic agents can relieve rigidity in the upper airway muscle, which may improve SDB in PD patients. A clinical have found that long-acting levodopa (controlled-release Sinemet) could alleviate OSA in the later half of the night in PD patients [17]. Another multinational, double-blind, RCT demonstrated that transdermal rotigotine patch may significantly improve 10 individual Modified PDSS-2 items including respiratory problems and snoring in PD patients [18].

7.4.3 Continuous Positive Airway Pressure Therapy (CPAP)

CPAP is the treatment option for OSA in the general population; however, it has not been comprehensively explored in PD patients. Results from a study indicated that in patients with PD having OSA, when compared to those treated with placebo CPAP, therapeutic CPAP could successfully alleviate OSA as well as deepen sleep [19]. However, not all PD patients can tolerate CPAP, especially in the advanced PD patients, where reduced mobility can increase difficulties with adjusting the CPAP apparatus. Moreover, other factors interfering the use of CPAP include nasal obstruction, sinus infection, chronic mouth breathing, and lack of motivation. When CPAP is limited in advanced PD patients, behavioral treatment options should be considered.

References

1. Qasrawi S, Al Ismaili R, Pandi-Perumal SR, et al. Synopsis of Sleep Medicine. New York: Apple Academic Press, 2016.
2. Albers JA, Chand P, Anch AM. Multifactorial sleep disturbance in Parkinson's disease. Sleep Med. 2017;35:41–8. https://doi.org/10.1016/j.sleep.2017.03.026.
3. Maria B, Sophia S, Michalis M, et al. Sleep breathing disorders in patients with idiopathic Parkinson's disease. Respir Med. 2003;97:1151–7.
4. Chotinaiwattarakul W, Dayalu P, Chervin RD, Albin RL. Risk of sleep-disordered breathing in Parkinson's disease. Sleep Breath. 2011;15:471–8.
5. Crosta F, Desideri G, Marini C. Obstructive sleep apnea syndrome in Parkinson's disease and other parkinsonisms. Funct Neurol. 2017;32:137–41.
6. Trotti LM, Bliwise DL. No increased risk of obstructive sleep apnea in Parkinson's disease. Mov Disord. 2010;25:2246–9.

7. Nomura T, Inoue Y, Kobayashi M, et al. Characteristics of obstructive sleep apnea in patients with Parkinson's disease. J Neurol Sci. 2013;327:22–4.

8. Kaminska M, Lafontaine A, Kimoff RJ. The interaction between obstructive sleep apnea and Parkinson's disease: possible mechanisms and implications for cognitive function. Parkinsons Dis. 2015;2015:849472.

9. Chahine LM, Amara AW, Videnovic A. A systematic review of the literature on disorders of sleep and wakefulness in Parkinson's disease from 2005 to 2015. Sleep Med Rev. 2017;35:33–50.

10. Braak H, Del K, Rüb U, et al. Staging of brain pathology related to sporadic Parkinson's disease. Neurobiol Aging. 2003;24:197–211.

11. Shelgikar AV, Chervin R. Approach to and evaluation of sleep disorders. Contin (Minneap Minn). 2013;19:32–49.

12. Kurtis MM, Balestrino R, Rodriguez-Blazquez C, et al. A review of scales to evaluate sleep disturbances in movement disorders. Front Neurol. 2018;9:369.

13. Berry RB, Budhiraja R, Gottlieb DJ, et al. Rules for scoring respiratory events in sleep: update of the 2007 AASM manual for the scoring of sleep and associated events. J Clin Sleep Med. 2012;8:597–619.

14. Videnovic A. Management of sleep disorders in Parkinson's disease and multiple system atrophy. Mov Disord. 2017;32:659–68.

15. Epstein LJ, Kristo D, Strollo PJ, et al. Clinical guideline for the evaluation, management and long-term care of obstructive sleep apnea in adults. J Clin Sleep Med. 2009;5:263–76.

16. Liu CF, Wang T, Zhan SQ, et al. Management recommendations on sleep disturbance of patients with Parkinson's disease. Chin Med J. 2018;131:2976–85.

17. Gros P, Mery VP, Lafontaine AL, et al. Obstructive sleep apnea in Parkinson's disease patients: effect of Sinemet CR taken at bedtime. Sleep Breath. 2016;20:205–12.

18. Trenkwalder C, Kies B, Sa F, et al. Rotigotine effects on early morning motor function and sleep in Parkinson's disease: A double-blind, randomized, placebo-controlled study (RECOVER). Mov Disord. 2011;26:90–9.

19. Neikrug AB, Liu L, Avanzino JA, et al. Continuous positive airway pressure improves sleep and daytime sleepiness in patients with Parkinson disease and sleep apnea. Sleep. 2014;37:177–85.

Excessive Daytime Sleepiness

8

Yun Shen and Chun-Feng Liu

Abstract

Excessive daytime sleepiness (EDS), one of the most common sleep abnormalities, is associated with many motor and non-motor symptoms in patients with Parkinson's disease (PD). Its causes are multifactorial, and it is first necessary to identify and treat any possible factors causing EDS. Recent studies show that some non-pharmacologic (i.e., cognitive behavioral therapy, light therapy, repetitive transcranial magnetic stimulation) and pharmacologic (i.e., modafinil, methylphenidate, caffeine, istradefylline, sodium oxybate, atomoxetine) treatments may be effective in treating EDS in PD. Further investigations are required to determine the safety and efficacy of potential therapies and to develop novel treatment approaches for EDS in PD.

Keywords

Parkinson's disease · Excessive daytime sleepiness · Sleep disorders

In recent years, the non-motor symptoms of Parkinson's disease (PD), the second most common neurodegenerative disorder, have received increasing attention, one of which is excessive daytime sleepiness (EDS). In this review, we summarize recent studies on the epidemiology, etiology, clinical implications, associated features, and evaluation of EDS in PD. In addition, we review the efficacy of pharmacologic and non-pharmacologic treatments for EDS in PD.

Y. Shen · C.-F. Liu (✉)
Department of Neurology, The Second Affiliated Hospital of Soochow University,
Suzhou, China
e-mail: liuchunfeng@suda.edu.cn

© Springer Nature Singapore Pte Ltd. 2020
C.-F. Liu (ed.), *Sleep Disorders in Parkinson's Disease*,
https://doi.org/10.1007/978-981-15-2481-3_8

EDS is defined as an inability to maintain wakefulness and alertness during the major waking episodes of the day that results in periods of irrepressible need for sleep or unintended lapses into drowsiness or sleep [1]. EDS is a major health hazard in PD, affecting 21–76% of PD patients with an incidence of 6% per year [2–4]. The prevalence of EDS is higher in PD patients than in the general population, with subjective sleepiness in 34–54% of PD patients compared with 16–19% of controls [5, 6].

An important feature of EDS is the "sudden onset of sleep," which is when a patient suddenly falls asleep during periods of inactivity or low activity. Sudden onset of sleep is reported by 1–31% of PD patients, and most patients are not able to completely recall the event [7, 8]. Some PD patients also exhibit significant features of narcolepsy, including cataplexy and sleep-onset REM periods (SOREMPs) in the multiple sleep latency test (MSLT) [9]. Sudden onset of sleep contributes significantly to disease burden and negatively impacts quality of life [10], impairs daytime functioning, and is associated with motor vehicle crashes. However, many PD patients may not be aware of their sleepiness.

8.1   Clinical Implications

EDS in PD is not persistent, and its presence may fluctuate over time. In general, the proportion of PD patients with EDS increases over time with longer follow-up. A longitudinal study revealed a progressive increase in EDS prevalence from 4% at baseline to 41% after 8 years of follow-up [11].

EDS is associated with and influences other motor and non-motor symptoms of PD. Longitudinal studies report that the presence of EDS is associated with clinical variables such as male gender, poorer nighttime sleep, cognitive impairment, autonomic dysfunction, hallucinations, depression, anxiety, probable behavior disorder, advanced disease, the postural-instability-gait-difficulty motor phenotype, less severe dyskinesias, dosage of dopamine agonists, and use of antihypertensives [12–14]. A longitudinal study shows that predictors of incident development of EDS include autonomic dysfunction, anxiety, and cerebrospinal fluid phosphorylated tau/total tau ratio [14].

Clinicians have also noted the impact of mood symptoms on EDS in PD. A recent review article reports a significant positive correlation between depression and EDS and a weak correlation between anxiety and EDS in PD patients. The magnitude of the correlation depended on how EDS was measured; it was medium when EDS was subjectively measured and small when EDS was objectively measured [15].

PD patients with EDS exhibit alterations in brain structure and function (e.g., brain volume, white matter integrity as indicated by fractional anisotropy, cerebral metabolism) [15]. It is impossible to determine whether EDS is a potential manifestation of more severe brainstem neurodegeneration.

Some evidence suggests that EDS predates the conversion of PD. Sleepy adults have a more than threefold increased risk of PD compared with non-sleepy adults (odds ratio, 3.3; 95% confidence interval [CI], 1.4–7.0; $p < 0.01$) [16]. Another study shows that EDS (i.e., an Epworth Sleepiness Scale [ESS] score >8 at the time of rapid eye movement [REM] sleep behavior disorder [RBD] diagnosis) predicts more rapid conversion to parkinsonism and dementia in idiopathic RBD (iRBD) patients [17]. Another study shows that EDS (i.e., ESS score ≥14) is significantly associated with an increased risk of developing PD in iRBD patients (adjusted hazard ratio, 3.6; 95% CI, 1.6–7.9; $p < 0.01$) [18]. By contrast, a prospective follow-up study assessing a large cohort of patients with iRBD and controls found no difference in baseline ESS score between those who eventually converted and those who remained disease-free [19]. This discrepancy between studies may result from differences in sample size, follow-up period, cutoff values for ESS score, and conversion time from RBD diagnosis or onset.

EDS may also be independently associated with risk of cognitive decline. Among 4894 elderly people, those who felt sleepy during the daytime had an increased risk of cognitive decline 8 years later [20].

8.2 Etiology

The etiology of EDS in PD is multifactorial. First, EDS may involve alterations in pathophysiological mechanisms involved in the regulation of sleep and wakefulness. In the brainstem, neurodegeneration within ascending arousal systems controls neurotransmission across several neuronal nuclei such as the noradrenergic locus coeruleus, noradrenergic dorsal motor nucleus of the vagus nerve, serotonergic dorsal raphe nucleus, histaminergic tuberomammillary nucleus, and dopaminergic areas. In particular, adenosine is a neurotransmitter that promotes non-REM sleep, and cholinergic neurons in laterodorsal tegmental and pedunculopontine tegmental nuclei promote REM sleep. Because PD progression may co-occur with the degeneration of neurons controlling wakefulness and sleep, it could lead to sleep disorders including EDS [9, 21].

Second, EDS could be an adverse outcome of dopaminergic therapy. Several studies show that dopaminergic agents (e.g., levodopa) and agonists (e.g., pramipexole, ropinirole, rotigotine) cause somnolence [22–24]. PD patients taking a dopamine agonist are sleepier than those treated with levodopa alone [25–27]. Combination therapy with levodopa and a dopamine agonist is associated with the highest risk of EDS [28]. Also, the influence of dopaminergic therapy on EDS is dose dependent [29, 30], and some investigators believe that PD patients who take high doses of dopaminergic therapy are prone to irresistible sleep attacks [31].

Third, EDS may be linked to poor (i.e., non-restorative) nocturnal sleep. Polysomnographic studies show that PD patients have significantly shorter total sleep time, lower sleep efficiency, and sleep architectural changes [32]. Many coexistent primary sleep disorders (e.g., restless legs syndrome (RLS), periodic limb movement disorder, RBD), motor disturbances (e.g., nocturnal akinesia, bradykinesia, rest tremor, inability to turn over in bed), and other non-motor symptoms (i.e., pain, depression, nocturia, hallucinations, temperature dysregulation due to dysautonomia) could also lead to sleep fragmentation, which in turn could result in EDS [33–35]. In particular, the presence of RBD might be associated with greater sleepiness in PD, as some studies report that PD patients with EDS have a higher rate of RBD than those without EDS [36] and that PD patients with probable RBD experience a higher level of sleepiness than those without RBD [37]. It is unclear whether RLS directly contributes to EDS. One study shows no difference in subjective sleepiness between PD patients with and without RLS. However, the SLEEMSA study reports that RLS predicts EDS in PD [6, 38]. Furthermore, although sleep-disordered breathing may play a role in EDS, its overall contribution may be limited [39, 40].

In addition, other factors such as genes; the sleep environment; use of antihypertensive medications, benzodiazepines, antipsychotics, and certain antidepressants (e.g., serotonin-selective reuptake inhibitors, MAO-I); [41] hypocretin (orexin) cell loss; [42] and circadian rhythm abnormalities [43] may also contribute to EDS in PD. Therefore, further studies are required in this area.

8.3   Evaluation of EDS in PD

PD patients should undergo thorough sleep evaluation consisting of solicitation of the chief complaint; detailed social, family, medical, psychiatric, and sleep history; physical examination; and, if necessary, objective sleep testing with polysomnography and the MSLT.

8.3.1   Subjective Assessment

Clinicians should understand the clinical presentation of sleepiness by PD patients. Patients might complain of both daytime sleepiness and disturbed nocturnal sleep and sometimes have associated complaints such as daytime fatigue, lack of concentration, and lack of symptom relief after additional sleep. It is important to distinguish sleepiness from fatigue, as there is significant overlap between the two symptoms [44]. Fatigue is a physical or psychological feeling that can be confounded with EDS. Fatigued patients may describe themselves as feeling tired or having a lack of energy, but they do not fall asleep when sedentary [45]. Interviewing other people who are familiar with the patient could help provide more information than can be obtained directly from the patient.

Subjective scales have some advantages in terms of their ease of administration and ability to incorporate patient insight into the degree of the problem. The ESS is a scale that is commonly used to determine the severity of sleepiness during a given period of time. Many studies use a score of 10 as a cutoff to identify sleepiness. Other useful questionnaires for assessing sleepiness include the Stanford Sleepiness Scale, Pittsburgh Sleep Quality Index, Scales for Outcomes in Parkinson's Disease-SLEEP-Daytime Sleepiness, and PD Sleep Scale. However, clinical impression of sleepiness and results of sleep questionnaires might be insufficient evidence for medical concern [46].

8.3.2 Objective Assessment

To reduce bias and the potential impact of confounding factors, objective measures may also be appropriate. Polysomnography can identify underlying sleep disorders such as obstructive sleep apnea, insomnia, and RBD that cause night sleep fragmentation and can provide indirect evidence of EDS. Also, standardized tests for assessing EDS are the MSLT and Maintenance of Wakefulness Test. The MSLT assesses ability to fall asleep, whereas the Maintenance of Wakefulness Test assesses ability to remain awake. These two tests are not routinely used to evaluate sleepiness in PD. One study found a high frequency of self-reported EDS in PD patients despite that many patients do not exhibit short sleep latency in the MSLT [47]. However, when PD patients exhibit narcolepsy-like behavior, the MSLT could demonstrate mean sleep latency and SOREMP for differentially diagnosing narcolepsy. Furthermore, a 24-h continuous sleep recording or an actigraphic recording of at least 1 week can also be used to diagnose EDS in PD.

We believe that the appropriate selection of a subjective or objective assessment of EDS in PD is very important. Currently, there is no commonly accepted clinically useful method for diagnosing EDS in PD. Considering subjective assessments, each scale has its own advantages and disadvantages, and different studies adopt different measures of EDS, which can result in varying conclusions. Most studies compare the results of sleep questionnaires to the extensively used ESS, as it is important to understand the sensitivity, specificity, and application ranges of different criteria. Considering objective assessments, evidence of their value is low. Routine detection methods have not found meaningful results, although perhaps changes in the duration and times of testing could reveal different findings. There is also a need for more standardized assessment of EDS in PD. Therefore, objective evaluation in conjunction with subjective assessment may be the best approach to diagnosing EDS in PD.

8.4   Management of EDS in PD

There are few specific guidelines for treating EDS and no studies on the treatment of sudden onset of sleep in PD. The treatment of EDS is complex because of its heterogeneous causes in PD patients. Treatment must be individualized and directed at the underlying causes, if known [48]. Considering the available evidence, we discuss some implications for clinical practice. The efficacy level of "clinically useful" means that evidence available is sufficient to conclude that the intervention provides clinical benefit for a given situation. "Possibly useful" means that the available evidence is suggestive but insufficient to conclude that the intervention provides clinical benefit in a given situation. "Investigational" means that the available evidence is insufficient to support the use of the intervention in clinical practice, although further study is warranted. We found no "unlikely useful" or "not useful" management approaches [49].

In addition to PD-related motor disabilities, EDS and sudden onset of sleep while driving are critical factors for traffic safety. Patients should be warned not to drive if they doze in unusual circumstances [2], especially when their ESS score is ≥7 [2]. It is also necessary to identify and treat any possible sleep disorders, for example, some sleep disorders could disrupt nocturnal sleep and some possible drugs causing hypersomnia, such as antidepressants, antipsychotics, or sedatives. The European Federation of Neurological Societies and Movement Disorder Society-European Section recommend that managing EDS in PD patients should involve assessing nocturnal sleep disturbances; improving nocturnal sleep by reducing akinesia, tremor, and urinary frequency; recommending the cessation of driving; reducing or discontinuing sedative drugs; and reducing the dosage of dopaminergic drugs (mainly dopamine agonists) or switching to other dopamine agonists [50].

Dopaminergic therapies could improve overnight sleep and be useful for treating RLS in PD and thereby improving EDS. However, dopamine agonist-associated sleep abnormalities, including sleep attacks, should be considered potential dose-dependent risks of dopamine and combination therapy with levodopa and dopamine agonists. Dosage reduction, monotherapy, or discontinuation in these patients could be helpful and could be replaced by selegeline, amantadine, or entacapone, which have no effects on EDS [44, 51] and may even reduce or resolve EDS [4, 28, 51–53]. Although clonazepam is the mainstay of treatment for RBD, given its associated risks, benzodiazepine use should generally be avoided in PD patients with EDS. Furthermore, EDS occurs in nearly half of PD patients treated with clozapine [54].

Other primary sleep disorders that might cause EDS should be carefully assessed using polysomnography and treated appropriately. For example, continuous positive airway pressure treatment improves subjective and objective EDS in PD patients with obstructive sleep apnea by reducing apnea events, improving oxygen saturation, and deepening sleep [55]. *Efficacy conclusion:* Clinically useful. Identifying and treating primary sleep disorders is necessary and must be completed before any treatment.

8.4.1   Non-pharmacologic Therapies for EDS

Because drug therapies have the potential for adverse side effects, non-pharmacologic treatment approaches offer a promising alternative for preventing and managing EDS in PD.

Cognitive behavioral therapy Cognitive behavioral therapy for insomnia (CBT-I) is extensively used to treat insomnia in non-PD populations. It consists of behavioral and psychological approaches to teaching patients how to change their dysfunctional behaviors and thinking patterns. One small study found that the Insomnia Severity Index, PD Sleep Scale, and examiner-reported clinical global impression improved in PD patients who received CBT-I combined with light therapy [56]. Therefore, in accordance with CBT-I, clinicians could recommend that patients strictly follow sleep hygiene rules such as having regular nap times and daytime physical activity and avoiding vigorous physical activity 3–4 h before sleeping [4, 57]. *Efficacy conclusion:* Investigational. CBT-I is simple to administer, but there remains insufficient evidence for its effective management of EDS in PD patients.

Light therapy Supplementary exposure to bright light (i.e., light therapy) has beneficial effects on sleep, depression, bradykinesia, rigidity, and dyskinesias in PD patients, as light activates the suprachiasmatic nucleus and is the most effective zeitgeber of the circadian timing system. A randomized, placebo-controlled clinical study found that bright light therapy twice daily in 1-h intervals for 14 days significantly reduced ESS in PD patients (mean [standard deviation], 15.8 [3.1] at baseline vs. 11.2 [3.3] after intervention). Possible reasons for this effect are that light therapy improves PD severity, daytime alertness, nighttime sleep quality, or sleep fragmentation by influencing the circadian system [58] and promoting dopamine release [59]. Because light therapy is noninvasive and has only mild and transient side effects, including headache, nausea, and hypomania, consideration of its use is warranted for the treatment of EDS in PD. However, there is a lack of consensus on the optimal parameters of light therapy for PD. *Efficacy conclusion:* Possibly useful. Light therapy may potentially be efficacious in preventing EDS, although future studies are required to determine its optimal timing, dosage, and treatment duration.

Repetitive transcranial magnetic stimulation Transcranial magnetic stimulation (TMS) is a noninvasive tool applied in different paradigms to obtain direct measures of cortical excitability. Repetitive TMS (rTMS) induces direct, trans-synaptic neuronal activation. High-frequency rTMS (>5 Hz) increases cortical excitability, whereas low-frequency rTMS (<1 Hz) has the opposite effect [60]. Combining rTMS with electroencephalography may be a useful approach to treating sleep disorders such as obstructive sleep apnea, RLS, narcolepsy, RBD, sleepwalking, sleep-wake disturbances after traumatic brain injury, and chronic insomnia [61, 62]. In PD patients, rTMS improves motor deficits (considering both UPDRS-III scores and gait parameters) [63]. One recent case report of a narcolepsy patient who received 25 sessions of high-frequency rTMS over the left dorsolateral prefrontal cortex demonstrated that rTMS might be a safe and effective alternative strategy for treating narcolepsy-like symptoms [64]. *Efficacy conclusion:* Investigational. No studies have yet focused on the efficacy of rTMS for treating EDS in PD. Future studies should seek to define optimal stimulation parameters, such as timing, duration, electrode placement, coil orientation, and physiological state of the patient.

8.4.2 Pharmacologic Therapies for EDS

If non-pharmacologic strategies do not improve EDS, drug therapies can be considered. There are few recommendations for the pharmacological management of EDS in PD, as few multi-center clinical trials have been conducted in this area [65, 66]. A Movement Disorder Society evidence-based medicine review concludes that there is insufficient data to recommend any specific drug for the long-term treatment of EDS in PD patients [67]. Limited data exist for the use of wakefulness-promoting agents such as modafinil and armodafinil or stimulants such as methylphenidate or dextroamphetamines.

Modafinil Modafinil, a medication approved by the US Food and Drug Administration to treat narcolepsy, is a wake-promoting agent and is indicated for most forms of EDS. A recent meta-analysis reports that modafinil effectively reduces ESS score, with an overall mean difference of 2.2 (95% CI, −3.9 to −0.6) and no significant heterogeneity among studies [68]. However, modafinil does not alter objective measures of sleepiness [69, 70]. In clinical practice, modafinil is given once a day in the morning on an empty stomach. The starting dose is usually 100 mg and can be increased slowly to 400 mg as needed. Modafinil is well tolerated in the treatment of EDS and has a low prevalence of side effects [68] such as headache, nausea, dry mouth, and anorexia [71]. However, for older PD patients, especially those with severe cardiovascular disease or other underlying cardiac abnormalities, the cardiovascular effects of modafinil, including elevated blood pressure and heart rate [72], are a concern. However, these side effects appear to be mild and decrease with dose reduction [67]. As alternatives to modafinil, other drugs that are generally well tolerated with a low prevalence of side effects could be considered. Although data on their efficacy is limited, it is reasonable to consider their use for treating EDS in PD in clinical practice. *Efficacy conclusion:* Possibly useful. There is insufficient evidence to draw conclusions about the efficacy and safety of modafinil for treating EDS in PD, although its use might be helpful in clinical practice.

Methylphenidate Methylphenidate, the piperazine derivative of amphetamine, increases the release and inhibits the reuptake of catecholamines, including dopamine and norepinephrine. Its effects may be mediated by the restoration of balance between dopamine and norepinephrine neurotransmitters. An open-label study reports that methylphenidate dramatically reduces EDS in PD patients, with high doses of methylphenidate improving motor and gait symptoms in the presence and absence of levodopa [73]. In clinical practice, methylphenidate is initially prescribed at 10 mg per day, with a recommended maximum dose of up to 80 mg per day. Possible adverse events related to methylphenidate therapy are reduced appetite, nausea, headache, insomnia, and psychosis [74]. PD patients can receive methylphenidate 2 weeks after discontinuation of monoamine oxidase inhibitors. *Efficacy conclusion:* Possibly useful. There is insufficient evidence to draw conclusions about the efficacy and safety of methylphenidate for treating EDS in PD, although its use might be helpful in clinical practice.

Caffeine Caffeine, an adenosine antagonist, reduces somnolence in the general population. A decade ago, it also attracted attention due to its potential neuroprotective effect. A meta-analysis reports that caffeine reduces the risk of PD (relative risk, 0.7; 95% CI, 0.6–0.8) [75]. In a long-term randomized controlled trial assessing the effects of caffeine on EDS in PD, patients given up to 200 mg caffeine twice a day for 6 weeks showed a non-significant reduction in ESS score (−1.7 points; 95% CI, −3.6–0.1), whereas clinical global impression of EDS improved in per protocol analysis [76]. In a more recent study, caffeine slightly improved EDS over the first 6 months, with the clinical effect lessening over time [77]. There are many potential explanations for this discrepancy between studies, including different study populations and trial durations. Caffeine could affect EDS or the sensation of alertness and is an inexpensive intervention that is well tolerated in most individuals. *Efficacy conclusion:* Investigational. The magnitude of the impact of caffeine on EDS in PD patients is unclear. It may be reasonable to try intermittent moderate doses of caffeine and repeat if improvement is observed.

Sodium oxybate   Sodium oxybate is used to treat cataplexy and EDS in narcolepsy and has been tested in PD patients. An open-label polysomnographic study reports that nocturnally administered sodium oxybate improves subjective sleepiness, sleep quality, and fatigue as well as slow wave sleep in PD patients [78]. Recently, a randomized, double-blind, placebo-controlled, crossover, phase IIA study reported that sodium oxybate is effective in treating EDS and nocturnal sleep disturbance with class I evidence. This study used both objective and subjective assessments and showed that sodium oxybate improved mean sleep latency, ESS score, and slow wave sleep duration [79]. Sodium oxybate should be taken in the evening and once again during the night. Its side effects are nausea, insomnia, headache, dizziness, vomiting, weight loss, psychiatric complications, and sleep apnea [80]. It increases apnea-hypopnea index in PD patients [78] and induces de novo obstructive sleep apnea and parasomnia [79]. *Efficacy conclusion:* Possibly useful. Evidence suggests that sodium oxybate may be efficacious for treating EDS in PD. However, stringent patient monitoring and larger follow-up trials are warranted.

Istradefylline   A single-center, open-label study reports that istradefylline, a selective adenosine A2A receptor antagonist, significantly improved EDS 2 and 3 months after PD patients received 20–40 mg/day istradefylline once daily in the morning. The underlying mechanism may be that istradefylline enhances alertness while having no negative impact on sleep [81]. *Efficacy conclusion:* Investigational. The use of istradefylline might be helpful in clinical practice, although further studies are warranted.

Atomoxetine   Atomoxetine, a selective norepinephrine reuptake inhibitor, has been shown to be beneficial for EDS in PD patients, possibly through alerting effects on norepinephrine neurons in the locus ceruleus. Constipation and insomnia are the most common adverse events. *Efficacy conclusion:* Investigational. The use of atomoxetine might be helpful in clinical practice, although further studies are warranted.

8.5   Conclusions

EDS is common in the PD population and can have an immensely negative impact on quality of life. Its causes are multifactorial, which complicates its treatment. More and larger studies are needed to demonstrate the efficacy and safety of pharmacologic and non-pharmacologic treatments for EDS in PD. Furthermore, efforts should focus on planning and executing clinical trials to develop novel treatment approaches.

8.6   Case Report

A 69-year-old woman with idiopathic PD (Hoehn and Yahr stage 3) was admitted to the Department of Neurology on December 8, 2018. She complained of excessive daytime sleepiness (EDS) for the last 6 months.

She was initially diagnosed with PD in July 2013, with resting tremor of her right hand and slowing of motor functions. These years, she has been a mild progression. On December 8, 2018, her motor examination showed moderate bilateral cogwheel rigidity and resting tremor of the upper extremities. It was something wrong with her gait and balance. Her medications included Madopar 125 mg TID, pramipexole 0.25 mg TID, amantadine 100 mg BID, and Selegiline 5 mg daily.

The score of Epworth Sleepiness Scale (ESS) was 13. She refused to use any stimulants drugs. We suggested she received light therapy (bright light therapy, 10,000 lx, in the morning (9–11 AM) and afternoon (5–7 PM) daily, 2 weeks).

There were no adverse events associated with light therapy. On repeat examination her motor symptoms were stable and unchanged. The score of Epworth Sleepiness Scale (ESS) was 9. She also found the nocturnal sleep was better, especially initiate sleep and sleep fragmentation.

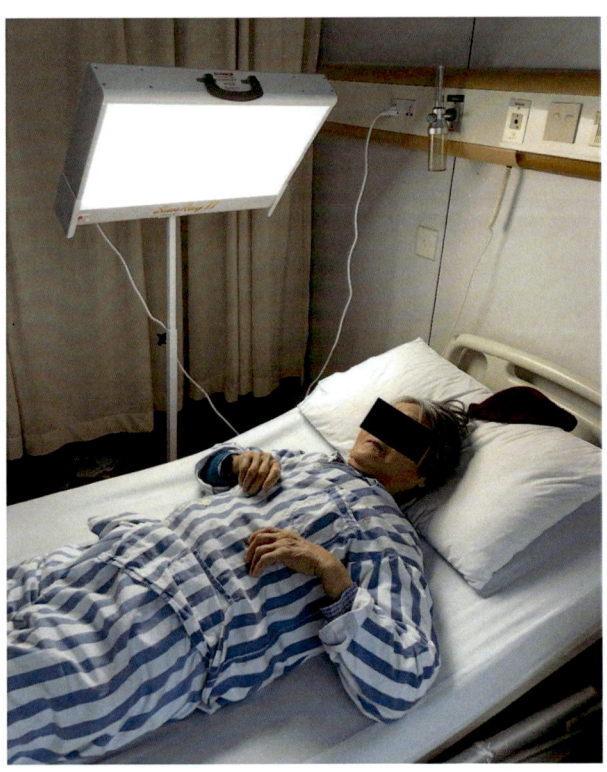

References

1. International Classification of Sleep Disorders III. American Academy of sleep medicine. Darien: IL; 2014.
2. Hobson DE, Lang AE, Martin WR, Razmy A, Rivest J, Fleming J. Excessive daytime sleepiness and sudden-onset sleep in Parkinson disease: a survey by the Canadian movement disorders group. JAMA. 2002;287(4):455–63. https://doi.org/10.1001/jama.287.4.455.
3. Falup-Pecurariu C, Diaconu S. Sleep dysfunction in Parkinson's Disease. Int Rev Neurobiol. 2017;133:719–42. https://doi.org/10.1016/bs.irn.2017.05.033.
4. Loddo G, Calandra-Buonaura G, Sambati L, Giannini G, Cecere A, Cortelli P, et al. The treatment of sleep disorders in Parkinson's disease: from research to clinical practice. Front Neurol. 2017;8:42. https://doi.org/10.3389/fneur.2017.00042.
5. Stavitsky K, Saurman JL, McNamara P, Cronin-Golomb A. Sleep in Parkinson's disease: a comparison of actigraphy and subjective measures. Parkinsonism Relat Disord. 2010;16(4):280–3. https://doi.org/10.1016/j.parkreldis.2010.02.001.
6. Chahine LM, Amara AW, Videnovic A. A systematic review of the literature on disorders of sleep and wakefulness in Parkinson's disease from 2005 to 2015. Sleep Med Rev. 2017;35:33–50. https://doi.org/10.1016/j.smrv.2016.08.001.
7. Ghorayeb I, Loundou A, Auquier P, Dauvilliers Y, Bioulac B, Tison F. A nationwide survey of excessive daytime sleepiness in Parkinson's disease in France. Mov Disord. 2007;22(11):1567–72. https://doi.org/10.1002/mds.21541.
8. Montastruc JL, Brefel-Courbon C, Senard JM, Bagheri H, Ferreira J, Rascol O, et al. Sleep attacks and antiparkinsonian drugs: a pilot prospective pharmacoepidemiologic study. Clin Neuropharmacol. 2001;24(3):181–3. https://doi.org/10.1097/00002826-200105000-00013.
9. Bliwise DL, Trotti LM, Juncos JJ, Factor SA, Freeman A, Rye DB. Daytime REM sleep in Parkinson's disease. Parkinsonism Relat Disord. 2013;19(1):101–3. https://doi.org/10.1016/j.parkreldis.2012.08.003.
10. Sobreira-Neto MA, Pena-Pereira MA, Sobreira EST, Chagas MHN, Fernandes RMF, Tumas V, et al. High frequency of sleep disorders in Parkinson's disease and its relationship with quality of life. Eur Neurol. 2017;78(5–6):330–7. https://doi.org/10.1159/000481939.
11. Gjerstad MD, Alves G, Wentzel-Larsen T, Aarsland D, Larsen JP. Excessive daytime sleepiness in Parkinson disease: is it the drugs or the disease? Neurology. 2006;67(5):853–8. https://doi.org/10.1212/01.wnl.0000233980.25978.9d.
12. Zhu K, van Hilten JJ, Marinus J. Course and risk factors for excessive daytime sleepiness in Parkinson's disease. Parkinsonism Relat Disord. 2016;24:34–40. https://doi.org/10.1016/j.parkreldis.2016.01.020.
13. Tholfsen LK, Larsen JP, Schulz J, Tysnes OB, Gjerstad MD. Development of excessive daytime sleepiness in early Parkinson disease. Neurology. 2015;85(2):162–8. https://doi.org/10.1212/WNL.0000000000001737.
14. Amara AW, Chahine LM, Caspell-Garcia C, Long JD, Coffey C, Hogl B, et al. Longitudinal assessment of excessive daytime sleepiness in early Parkinson's disease. J Neurol Neurosurg Psychiatry. 2017;88(8):653–62. https://doi.org/10.1136/jnnp-2016-315023.
15. Wen MC, Chan LL, Tan LCS, Tan EK. Mood and neural correlates of excessive daytime sleepiness in Parkinson's disease. Acta Neurol Scand. 2017;136(2):84–96. https://doi.org/10.1111/ane.12704.
16. Abbott RD, Ross GW, White LR, Tanner CM, Masaki KH, Nelson JS, et al. Excessive daytime sleepiness and subsequent development of Parkinson disease. Neurology. 2005;65(9):1442–6. https://doi.org/10.1212/01.wnl.0000183056.89590.0d.
17. Arnulf I, Neutel D, Herlin B, Golmard JL, Leu-Semenescu S, Cochen de Cock V, et al. Sleepiness in idiopathic REM sleep behavior disorder and Parkinson Disease. Sleep. 2015;38(10):1529–35. https://doi.org/10.5665/sleep.5040.

18. Zhou J, Zhang J, Lam SP, Chan JW, Mok V, Chan A, et al. Excessive daytime sleepiness predicts neurodegeneration in idiopathic REM sleep behavior disorder. Sleep. 2017;40(5):zsx041. https://doi.org/10.1093/sleep/zsx041.

19. Postuma RB, Gagnon JF, Pelletier A, Montplaisir JY. Insomnia and somnolence in idiopathic RBD: a prospective cohort study. NPJ Parkinsons Dis. 2017;3:9. https://doi.org/10.1038/s41531-017-0011-7.

20. Jaussent I, Bouyer J, Ancelin ML, Berr C, Foubert-Samier A, Ritchie K, et al. Excessive sleepiness is predictive of cognitive decline in the elderly. Sleep. 2012;35(9):1201–7. https://doi.org/10.5665/sleep.2070.

21. Sulzer D, Surmeier DJ. Neuronal vulnerability, pathogenesis, and Parkinson's disease. Mov Disord. 2013;28(6):715–24. https://doi.org/10.1002/mds.25187.

22. Parkinson SG. Pramipexole vs levodopa as initial treatment for Parkinson disease: a randomized controlled trial. Parkinson Study Group JAMA. 2000;284(15):1931–8. https://doi.org/10.1001/jama.284.15.1931.

23. Tanner CM. Dopamine agonists in early therapy for Parkinson disease: promise and problems. JAMA. 2000;284(15):1971–3. https://doi.org/10.1001/jama.284.15.1971.

24. Comella CL. Daytime sleepiness, agonist therapy, and driving in Parkinson disease. JAMA. 2002;287(4):509–11. https://doi.org/10.1001/jama.287.4.509.

25. Ondo WG, Dat Vuong K, Khan H, Atassi F, Kwak C, Jankovic J. Daytime sleepiness and other sleep disorders in Parkinson's disease. Neurology. 2001;57(8):1392–6. https://doi.org/10.1212/WNL.57.8.1392.

26. Avorn J, Schneeweiss S, Sudarsky LR, Benner J, Kiyota Y, Levin R, et al. Sudden uncontrollable somnolence and medication use in Parkinson disease. Arch Neurol. 2005;62(8):1242–8. https://doi.org/10.1001/archneur.62.8.1242.

27. O'Suilleabhain PE, Dewey RB Jr. Contributions of dopaminergic drugs and disease severity to daytime sleepiness in Parkinson disease. Arch Neurol. 2002;59(6):986–9. https://doi.org/10.1001/archneur.59.6.986.

28. Paus S, Brecht HM, Koster J, Seeger G, Klockgether T, Wullner U. Sleep attacks, daytime sleepiness, and dopamine agonists in Parkinson's disease. Mov Disord. 2003;18(6):659–67. https://doi.org/10.1002/mds.10417.

29. Arnulf I, Konofal E, Merino-Andreu M, Houeto JL, Mesnage V, Welter ML, et al. Parkinson's disease and sleepiness: an integral part of PD. Neurology. 2002;58(7):1019–24. https://doi.org/10.1212/WNL.58.7.1019.

30. Hauser RA, Gauger L, Anderson WM, Zesiewicz TA. Pramipexole-induced somnolence and episodes of daytime sleep. Mov Disord. 2000;15(4):658–63. https://doi.org/10.1002/1531-8257(200007)15:4%3C658::AID-MDS1009%3E3.0.CO;2-N.

31. Frucht S, Rogers JD, Greene PE, Gordon MF, Fahn S. Falling asleep at the wheel: motor vehicle mishaps in persons taking pramipexole and ropinirole. Neurology. 1999;52(9):1908–10. https://doi.org/10.1212/WNL.52.9.1908.

32. Shpirer I, Miniovitz A, Klein C, Goldstein R, Prokhorov T, Theitler J, et al. Excessive daytime sleepiness in patients with Parkinson's disease: a polysomnography study. Mov Disord. 2006;21(9):1432–8. https://doi.org/10.1002/mds.21002.

33. Bhat S, Chokroverty S. Hypersomnia in neurodegenerative diseases. Sleep Med Clin. 2017;12(3):443–60. https://doi.org/10.1016/j.jsmc.2017.03.017.

34. Lees AJ, Blackburn NA, Campbell VL. The nighttime problems of Parkinson's disease. Clin Neuropharmacol. 1988;11(6):512–9.

35. Ferreira JJ, Desboeuf K, Galitzky M, Thalamas C, Brefel-Courbon C, Fabre N, et al. Sleep disruption, daytime somnolence and 'sleep attacks' in Parkinson's disease: a clinical survey in PD patients and age-matched healthy volunteers. Eur J Neurol. 2006;13(3):209–14. https://doi.org/10.1111/j.1468-1331.2006.01262.x.

36. Simuni T, Caspell-Garcia C, Coffey C, Chahine LM, Lasch S, Oertel WH, et al. Correlates of excessive daytime sleepiness in de novo Parkinson's disease: a case control study. Mov Disord. 2015;30(10):1371–81. https://doi.org/10.1002/mds.26248.

37. Rolinski M, Szewczyk-Krolikowski K, Tomlinson PR, Nithi K, Talbot K, Ben-Shlomo Y, et al. REM sleep behaviour disorder is associated with worse quality of life and other non-motor features in early Parkinson's disease. J Neurol Neurosurg Psychiatry. 2014;85(5):560–6. https://doi.org/10.1136/jnnp-2013-306104.

38. Moreno-Lopez C, Santamaria J, Salamero M, Del Sorbo F, Albanese A, Pellecchia MT, et al. Excessive daytime sleepiness in multiple system atrophy (SLEEMSA study). Arch Neurol. 2011;68(2):223–30. https://doi.org/10.1001/archneurol.2010.359.

39. Yeh NC, Tien KJ, Yang CM, Wang JJ, Weng SF. Increased risk of Parkinson's disease in patients with obstructive sleep apnea: a population-based, propensity score-matched, longitudinal follow-up study. Medicine (Baltimore). 2016;95(2):e2293. https://doi.org/10.1097/MD.0000000000002293.

40. Cochen De Cock V, Abouda M, Leu S, Oudiette D, Roze E, Vidailhet M, et al. Is obstructive sleep apnea a problem in Parkinson's disease? Sleep Med. 2010;11(3):247–52. https://doi.org/10.1016/j.sleep.2009.05.008.

41. Waters F, Faulkner D, Naik N, Rock D. Effects of polypharmacy on sleep in psychiatric inpatients. Schizophr Res. 2012;139(1–3):225–8. https://doi.org/10.1016/j.schres.2012.05.013.

42. Thannickal TC, Lai YY, Siegel JM. Hypocretin (orexin) cell loss in Parkinson's disease. Brain. 2007;130(Pt 6):1586–95. https://doi.org/10.1093/brain/awm097.

43. Videnovic A, Noble C, Reid KJ, Peng J, Turek FW, Marconi A, et al. Circadian melatonin rhythm and excessive daytime sleepiness in Parkinson disease. JAMA Neurol. 2014;71(4):463–9. https://doi.org/10.1001/jamaneurol.2013.6239.

44. Valko PO, Waldvogel D, Weller M, Bassetti CL, Held U, Baumann CR. Fatigue and excessive daytime sleepiness in idiopathic Parkinson's disease differently correlate with motor symptoms, depression and dopaminergic treatment. Eur J Neurol. 2010;17(12):1428–36. https://doi.org/10.1111/j.1468-1331.2010.03063.x.

45. Monderer R, Ahmed IM, Thorpy M. Evaluation of the sleepy patient: differential diagnosis. Sleep Med Clin. 2017;12(3):301–12. https://doi.org/10.1016/j.jsmc.2017.03.006.

46. Murray BJ. Subjective and objective assessment of Hypersomnolence. Sleep Med Clin. 2017;12(3):313–22. https://doi.org/10.1016/j.jsmc.2017.03.007.

47. Cochen De Cock V, Bayard S, Jaussent I, Charif M, Grini M, Langenier MC, et al. Daytime sleepiness in Parkinson's disease: a reappraisal. PLoS One. 2014;9(9):e107278. https://doi.org/10.1371/journal.pone.0107278.

48. Takenoshita S, Nishino S. Pharmacologic management of excessive daytime sleepiness. Sleep Med Clin. 2017;12(3):461–78. https://doi.org/10.1016/j.jsmc.2017.03.019.

49. Fox SH, Katzenschlager R, Lim SY, Ravina B, Seppi K, Coelho M, et al. The Movement Disorder Society evidence-based medicine review update: treatments for the motor symptoms of Parkinson's disease. Mov Disord. 2011;26(Suppl 3):S2–41. https://doi.org/10.1002/mds.23829.

50. Ferreira JJ, Katzenschlager R, Bloem BR, Bonuccelli U, Burn D, Deuschl G, et al. Summary of the recommendations of the EFNS/MDS-ES review on therapeutic management of Parkinson's disease. Eur J Neurol. 2013;20(1):5–15. https://doi.org/10.1111/j.1468-1331.2012.03866.x.

51. Diederich NJ, McIntyre DJ. Sleep disorders in Parkinson's disease: many causes, few therapeutic options. J Neurol Sci. 2012;314(1–2):12–9. https://doi.org/10.1016/j.jns.2011.10.025.

52. Stevens S, Cormella CL, Stepanski EJ. Daytime sleepiness and alertness in patients with Parkinson disease. Sleep. 2004;27(5):967–72. https://doi.org/10.1093/sleep/27.5.967.

53. Razmy A, Lang AE, Shapiro CM. Predictors of impaired daytime sleep and wakefulness in patients with Parkinson disease treated with older (ergot) vs newer (nonergot) dopamine agonists. Arch Neurol. 2004;61(1):97–102. https://doi.org/10.1001/archneur.61.1.97.

54. Amara AW, Chahine LM, Videnovic A. Treatment of sleep dysfunction in Parkinson's disease. Curr Treat Options Neurol. 2017;19(7):26. https://doi.org/10.1007/s11940-017-0461-6.

55. Neikrug AB, Liu L, Avanzino JA, Maglione JE, Natarajan L, Bradley L, et al. Continuous positive airway pressure improves sleep and daytime sleepiness in patients with Parkinson disease and sleep apnea. Sleep. 2014;37(1):177–85. https://doi.org/10.5665/sleep.3332.

56. Rios Romenets S, Creti L, Fichten C, Bailes S, Libman E, Pelletier A, et al. Doxepin and cognitive behavioural therapy for insomnia in patients with Parkinson's disease—a randomized study. Parkinsonism Relat Disord. 2013;19(7):670–5. https://doi.org/10.1016/j.parkreldis.2013.03.003.

57. Leroi I, Baker P, Kehoe P, Daniel E, Byrne EJ. A pilot randomized controlled trial of sleep therapy in Parkinson's disease: effect on patients and caregivers. Int J Geriatr Psychiatry. 2010;25(10):1073–9. https://doi.org/10.1002/gps.2472.

58. Videnovic A, Klerman EB, Wang W, Marconi A, Kuhta T, Zee PC. Timed light therapy for sleep and daytime sleepiness associated with Parkinson disease: a randomized clinical trial. JAMA Neurol. 2017;74(4):411–8. https://doi.org/10.1001/jamaneurol.2016.5192.

59. Li Z, Tian T. Light therapy promoting dopamine release by stimulating retina in Parkinson Disease. JAMA Neurol. 2017;74(10):1267–8. https://doi.org/10.1001/jamaneurol.2017.1906.

60. Medeiros LF, de Souza IC, Vidor LP, de Souza A, Deitos A, Volz MS, et al. Neurobiological effects of transcranial direct current stimulation: a review. Front Psych. 2012;3:110. https://doi.org/10.3389/fpsyt.2012.00110.

61. Civardi C, Collini A, Monaco F, Cantello R. Applications of transcranial magnetic stimulation in sleep medicine. Sleep Med Rev. 2009;13(1):35–46. https://doi.org/10.1016/j.smrv.2008.04.001.

62. Nardone R, Holler Y, Brigo F, Tezzon F, Golaszewski S, Trinka E. Transcranial magnetic stimulation and sleep disorders: pathophysiologic insights. Sleep Med. 2013;14(11):1047–58. https://doi.org/10.1016/j.sleep.2013.04.025.

63. Goodwill AM, Lum JAG, Hendy AM, Muthalib M, Johnson L, Albein-Urios N, et al. Using non-invasive transcranial stimulation to improve motor and cognitive function in Parkinson's disease: a systematic review and meta-analysis. Sci Rep. 2017;7(1):14840. https://doi.org/10.1038/s41598-017-13260-z.

64. Lai JB, Han MM, Xu Y, Hu SH. Effective treatment of narcolepsy-like symptoms with high-frequency repetitive transcranial magnetic stimulation: a case report. Medicine (Baltimore). 2017;96(46):e8645. https://doi.org/10.1097/MD.0000000000008645.

65. National Collaborating Centre for Chronic Conditions. Parkinson's disease: National clinical guideline for diagnosis and management in primary and secondary care. London: Royal College of Physicians; 2006.

66. Zesiewicz TA, Sullivan KL, Arnulf I, Chaudhuri KR, Morgan JC, Gronseth GS, et al. Practice parameter: treatment of nonmotor symptoms of Parkinson disease: report of the quality standards Subcommittee of the American Academy of Neurology. Neurology. 2010;74(11):924–31. https://doi.org/10.1212/WNL.0b013e3181d55f24.

67. Seppi K, Weintraub D, Coelho M, Perez-Lloret S, Fox SH, Katzenschlager R, et al. The Movement Disorder Society evidence-based medicine review update: treatments for the nonmotor symptoms of Parkinson's disease. Mov Disord. 2011;26(Suppl 3):S42–80. https://doi.org/10.1002/mds.23884.

68. Rodrigues TM, Castro Caldas A, Ferreira JJ. Pharmacological interventions for daytime sleepiness and sleep disorders in Parkinson's disease: systematic review and meta-analysis. Parkinsonism Relat Disord. 2016;27:25–34. https://doi.org/10.1016/j.parkreldis.2016.03.002.

69. Hogl B, Saletu M, Brandauer E, Glatzl S, Frauscher B, Seppi K, et al. Modafinil for the treatment of daytime sleepiness in Parkinson's disease: a double-blind, randomized, crossover, placebo-controlled polygraphic trial. Sleep. 2002;25(8):905–9.

70. Ondo WG, Fayle R, Atassi F, Jankovic J. Modafinil for daytime somnolence in Parkinson's disease: double blind, placebo controlled parallel trial. J Neurol Neurosurg Psychiatry. 2005;76(12):1636–9. https://doi.org/10.1136/jnnp.2005.065870.

71. Roth T, Schwartz JR, Hirshkowitz M, Erman MK, Dayno JM, Arora S. Evaluation of the safety of modafinil for treatment of excessive sleepiness. J Clin Sleep Med. 2007;3(6):595–602.

72. Dolder CR, Davis LN, McKinsey J. Use of psychostimulants in patients with dementia. Ann Pharmacother. 2010;44(10):1624–32. https://doi.org/10.1345/aph.1P341.

73. Devos D, Krystkowiak P, Clement F, Dujardin K, Cottencin O, Waucquier N, et al. Improvement of gait by chronic, high doses of methylphenidate in patients with advanced Parkinson's disease. J Neurol Neurosurg Psychiatry. 2007;78(5):470–5. https://doi.org/10.1136/jnnp.2006.100016.
74. Leonard BE, McCartan D, White J, King DJ. Methylphenidate: a review of its neuropharmacological, neuropsychological and adverse clinical effects. Hum Psychopharmacol. 2004;19(3):151–80. https://doi.org/10.1002/hup.579.
75. Noyce AJ, Bestwick JP, Silveira-Moriyama L, Hawkes CH, Giovannoni G, Lees AJ, et al. Meta-analysis of early nonmotor features and risk factors for Parkinson disease. Ann Neurol. 2012;72(6):893–901. https://doi.org/10.1002/ana.23687.
76. Postuma RB, Lang AE, Munhoz RP, Charland K, Pelletier A, Moscovich M, et al. Caffeine for treatment of Parkinson disease: a randomized controlled trial. Neurology. 2012;79(7):651–8. https://doi.org/10.1212/WNL.0b013e318263570d.
77. Postuma RB, Anang J, Pelletier A, Joseph L, Moscovich M, Grimes D, et al. Caffeine as symptomatic treatment for Parkinson disease (cafe-PD): a randomized trial. Neurology. 2017;89(17):1795–803. https://doi.org/10.1212/WNL.0000000000004568.
78. Ondo WG, Perkins T, Swick T, Hull KL Jr, Jimencz JE, Garris TS, et al. Sodium oxybate for excessive daytime sleepiness in Parkinson disease: an open-label polysomnographic study. Arch Neurol. 2008;65(10):1337–40. https://doi.org/10.1001/archneur.65.10.1337.
79. Buchele F, Hackius M, Schreglmann SR, Omlor W, Werth E, Maric A, et al. Sodium oxybate for excessive daytime sleepiness and sleep disturbance in Parkinson disease: a randomized clinical trial. JAMA Neurol. 2018;75(1):114–8. https://doi.org/10.1001/jamaneurol.2017.3171.
80. Wang YG, Swick TJ, Carter LP, Thorpy MJ, Benowitz NL. Safety overview of postmarketing and clinical experience of sodium oxybate (Xyrem): abuse, misuse, dependence, and diversion. J Clin Sleep Med. 2009;5(4):365–71.
81. Suzuki K, Miyamoto M, Miyamoto T, Uchiyama T, Watanabe Y, Suzuki S, et al. Istradefylline improves daytime sleepiness in patients with Parkinson's disease: an open-label, 3-month study. J Neurol Sci. 2017;380:230–3. https://doi.org/10.1016/j.jns.2017.07.045.

Circadian Rhythms Disruption

9

Guo-Dong Huang and Ya-Li Wang

Abstract

Abnormal circadian rhythm is quite common in PD patients, such as sleep-wake cycles, motor symptoms fluctuation, endocrine changes, autonomic dysfunction, and so on. It may have a negative effect on life quality of the patients. In addition, the disrupted biorhythm may alter the anti-oxidative ability, the autophagy level, and the mitochondrial function and thus accelerate disease progression. Recent studies showed that biorhythm modification, such as light therapy and physical exercise, can improve the motor symptoms and delay disease progression. Based on this, neurological clinicians should get more attention to circadian dysfunction of PD, and the circadian therapy may be a new hopeful strategy for PD.

Keywords

Circadian rhythms · Parkinson's disease

Circadian rhythms are biological rhythms with a periodicity of approximately 24 h in humans. These rhythms influence many physiological and behavioral functions. The sleep-wake cycle is one of the most robust outputs of circadian timekeeping mechanism.

Endogenous circadian rhythms can be characterized by analyzing circadian markers. There are alterations of the endogenous circadian rhythmicity in PD [1]. For example, some studies reported decreased amplitudes of melatonin secretion,

G.-D. Huang
Department of Neurology, Brigham and Women's Hospital, Harvard Medical School, Boston, MA, USA

Y.-L. Wang (✉)
Department of Neurology, Suzhou Municipal Hospital (North), Nanjing Medical University, Suzhou, China

which was significantly lower in PD patients with EDS. PD patients are also found to have elevated cortisol levels and flattened expression rhythm of a major core clock gene, Bmal1. Ambulatory BP monitoring for 24 h reveals significant differences in the rhythm of non-dipping, the percent of nocturnal BP decrease, nighttime BP levels, and nocturnal decrease of HR between PD patients and controls. These changes do not appear related to disease severity and phenotype, supporting the hypothesis that these alterations may stem from intrinsic circadian dysregulation. Circadian disruption has been associated with neuropsychiatric disturbances in PD.

Circadian dysfunction is under-recognized in PD and likely influences not only sleep-wake cycles but also may affect mood, cognition and autonomic and motor functions. Circadian-based therapies, such as timed light exposure and melatonin, should be considered in PD patients with evidence of circadian dysregulation.

Longitudinal studies centered on circadian function in PD will be needed to answer this and similar questions that may position circadian system as a novel diagnostic and therapeutic target in PD.

9.1   The Hierarchical Timing System in Mammals

The term circadian comes from the Latin circa *diem*, meaning "around or approximately a day." The biological clocks tick in almost all the living organisms such as humans, plants, animals, fungi, and cyanobacteria to coordinate diverse and widespread physiological processes and behaviors across approximately 24 h [1], so called circadian clocks. Starting the first *period*, fruit fly mutants were screened by Seymour and Konpka in 1971 that displayed alterations in the normal 24-h cycle of pupal eclosion and locomotor activity [2]. The 2017 Nobel Prize in Physiology or Medicine was awarded to Drs. Hall, Rosbash, and Young for their discovery on the notion of transcription-translation feedback loop (TTFL) that controls circadian rhythms. The circadian oscillators allow for anticipation of external environment changes and coordination of the internal machinery (TTFL) and play modulatory roles in molecular and behavioral decisions [3]. Recently, a lot of evidences have proved that the misalignment of the circadian clock could lead to numerous human diseases such as physical (diabetes, obesity) and psychiatric (sleep, neurodegeneration) disorders [4].

The circadian clock is genetically encoded by a series of conservative molecular components located in each kind of cell (even blood cells) that generates internal timing of 24 h in the absence of external environment signals [5]. The hierarchical "master" circadian clocks (also called core oscillator) are located in the suprachiasmatic nucleus (SCN) of the hypothalamus [6]. The master clock is the only core oscillator that anticipates and receives external light signal from the retina and then synchronizes to the peripheral clocks in the solar day [7]. Both the SCN and peripheral tissues share the same clock molecular mechanisms to maintain the rhythms in neurons or tissue cells [8]. Besides, core oscillator in the SCN was formed by a highly unified around 20,000 neurons [9, 10]. The phase of peripheral tissue clocks can be synchronized by the "master" clock via metabolic, hormone, and temperature cues [10, 11].

The cell-autonomous molecular clocks throughout the brain and the body in mammals are generated by two transcription-translation feedback loops that function together to generate robust 24-h rhythms for the circadian genes (PER and CRY) and circadian-controlled downstream genes (CCGs). The core TTFL is driven by two positive activators (circadian locomotor output cycles kaput (CLOCK) and brain and muscle ARNT-like 1 (BMAL1)) and two negative repressors (period (PER) and cryptochrome (CRY)). CLOCK and BMAL are subunits of the basic helix-loop-helix-PAS (bHLH-PAS) transcription factor. These two proteins form heterodimer and drive the transcript of *Per* and *Cry* genes via the E-box (CACGTG) or E'-box (CANNTG) motifs in the promoter region, as well as circadian-controlled genes [12, 13]. PER and CRY proteins form a heterodimer in the cytoplasm and then translocate to the nucleus to interact with CLOCK:BMAL complex and inhibit the transcription activity. Busino and Siepka found the stability of PER and CRY proteins related to FBXL3, a component in E3 ubiquitin degradation pathways [14, 15]. The second TTFL is constituted by retinoid-related orphan receptors (RORa, b, c) and REV-ERBα/REV-ERBβ that drive the transcription of *Bmal1* gene's activation and repression through RORE-box motif, respectively [16, 17]. Ukai's work showed that the rhythmic changes of *Bmal1* transcription are not required to maintain the core feedback loop; however, Andrew Liu found that the delayed expression of *Cry1* induced by the second TTFL loop (ROR/REV) is crucial for proper circadian timing [18, 19]. Therefore, the presence of cooperative, interlocking feedback loops not only provides the robustness against external perturbations to maintain the body circadian timing but also helps the core oscillator to output circadian signals to optimize the perfect timing of each gene's expression for local physiology.

Recently, several chromatin remodelers have been found involved in circadian regulator machinery. The first evidence was reported by Crosio, which showed that nighttime light exposure causes rapid, transient induction of clock in SCN neurons due to the phosphorylation at Ser10 in Histone 3 [20]. Masao found that CLOCK shares homology with acetyl-coenzyme binding motifs within the histone acetyltransferases (HATs), which can remodel the chromatin by acetyltransferase on histone H3 [21]. A larger number of histone deacetylases (HDACs) have been found to counterbalance these histone acetyltransferases. Duong found that PER can recruit SIN3A, a scaffold for assembly of transcriptional inhibitory complexes, in which the PER complex thereby rhythmically delivers histone deacetylases to the *Per1* promoter, which repress *Per1* transcription [22]. In contrast, Naruse found that CRY1 protein associates with the SIN3B-HDAC1/2 complex, and Feng found that REV-ERBα recruits the NCoR-HDAC3 complex in a rhythmic manner to remodel the chromatin [23, 24].

9.2   Circadian Rhythms, Sleep, and Neurodegeneration Disease

Sleep is an important and conserved phenomenon that has been identified in animals ranging from worms, zebrafish, and octopus to birds and mammals [25]. Sleep remains a poorly understood process although it is ubiquitous. While sleep has been proposed to be important in a series of life process such as metabolism, immune function, synaptic homeostasis and neural plasticity, and learning and memory [26–30]. In humans, sleep can be measured by electroencephalography (EEG) and broadly categorized into rapid eye movement (REM) sleep and "light" non-REM (NREM) sleep versus "deep" NREM sleep due to different sleep status [31].

The sleep-wake cycle is one of the most robust outputs of circadian timekeeping. The most compelling evidence linking circadian clock and sleep comes from the genetic analysis of the human sleep disorders such as familial advanced sleep phase syndrome (FASPS) and delayed sleep phase syndrome (DSPS) [32–34] (Table 9.1).

Table 9.1 Mutations in circadian genes cause human sleep disorders

Gene	Function	Mutation	Phenotype	References
CK1δ	Kinase	T44A	FASPS Sleep onset: 18:12 ± 1.4 h Wake up: 04:06 ± 0.7 h	[35]
Clock	Transcription factor	3111C allele	"Night-owl" tendency 10–44 min delay in activity or sleep	[36]
Cryptochrome1	Transcriptional repressor	1657 + 3A > C	DSPS Melatonin onset delay 2–2.5 h	[33]
Cryptochrome2	Transcriptional repressor	A260T	FASPS Accelerate cry degradation by FBXL3 Melatonin onset: Exp 16:41 vs Ctr 20:50	[37]
Dec2	Transcriptional repressor	P385R	Total sleep period (6.25 vs 8.06 h) Decrease NREM and REM sleep	[38]
Dec2	Transcriptional repressor	Y362H	Decrease baseline sleep period Decrease sleep	[39]
Period2	Transcriptional repressor	S622G	FASPS Sleep onset: 19:30 h Wake up onset: 04:30 h	[32]
Period3	Transcriptional repressor	Haplotype	DSPS	[40]
Period3	Transcriptional repressor	P415A H417R	FASPS Winter sleep onset: 19:00 vs 23:07 Winter sleep offset: 03:00 vs 07:45 Summer sleep onset/offset: normal	[34]

FASP Familial advanced sleep phase syndrome, *DSPS* delayed sleep phase disorder, *Exp* experimental group, *Ctr* control group

Toh KL found a serine to glycine mutation (S662G) within the casein kinase I epsilon (CKIε) binding region of hPER2, which causes a 4-h advance of the sleep, temperature, and melatonin rhythms [32]. Later Xu from the same group identified a missense mutation (T44A) in casein kinase I delta (CK1δ) that resulted in early sleep times and early-morning awakening. (FASPS) [35]. Hirano identified a missense mutation (A260T) in the human cryptochrome 2 gene that co-segregates with FASPS [37]. Another interesting story was reported that patients who are carrier of a Period3 variant (P415A and H417R) only showed advanced sleep syndrome in winter, suggesting that PER3 may participate in modulating circadian rhythms and mood to adapt to the short photoperiods in winter [34]. By contrast with the advanced sleep syndrome, delayed sleep syndrome is characterized by delayed sleep initiation, melatonin onset, and trouble awakening in the morning. Tranditionally, it is thought that delayed sleep phenomenon may result from voluntary behaviors of most patients such as watching electronic devices or media. But recently, the first genetic report correlated with DSPS was reported in polymorphisms of hPer3 patients by Takashi [40]. And another evidence was found by Patke which is a dominant coding variation in *Cry1* (c.1657+3A>C) as a direct DSPS allele [33]. Both FASPS and DSPS patients showed altered phase of the consolidated sleep period. On the molecular level, *Cry1* variant resulted in the skipping of exon11 and truncation of the C terminus and enhanced the repressor activity of CRY and lengthens the timing of feedback onto the CLOCK:BMAL1 heterodimer and the circadian period [33]. And the A260T mutation in *Cry2* showed shortened circadian period because of the genetic variant that enhances the degradation of CRY2 by FBXL3. However, scientists found that these circadian gene mutations generally do not affect sleep quality, and all the patients have intact circadian entrainment. Secondly, the changes in their circadian period length do not lead to non-24-h sleep-wake disorder. Instead, core oscillator in the SCN can receive the external signal from the environment and maintain the machinery mechanism then leading to the consistent advance or delay of their sleep time in each day.

The circadian system has been reported to be correlated with age-related neurodegenerative diseases such as Alzheimer's disease (AD), Parkinson's disease (PD), and Huntington's disease (HD) during aging process [41]. The disruption of sleep status and the impairment of the circadian rhythms (melatonin level, heart rate, temperature cycle, and so on) are the most common and early signs of these neurodegenerative diseases. The animal models and clinical patients with AD, PD, and HD showed impaired circadian rhythms and sleep cycle [42–45]. So how does the dysfunction of the circadian clock correlate with these neurodegenerative diseases? Does the circadian clock represent more as a neuroprotector in pathological process or as an abettor to promote neurodegeneration?

Melatonin is a rhythmic secreted peptide hormone that is mainly produced by the pineal gland and also other tissues in the body. As mentioned above, the genetic mutations of core clock genes not only impair the phase of the circadian clock but also obviously shift the release of melatonin level in cerebrospinal fluid or serum. In mammals, melatonin is involved in the entrainment of circadian rhythm including sleep-wake timing, blood pressure regulation, and seasonal reproduction, and it is also an antioxidant to protect nuclear and mitochondrial DNA [46–48]. Results showed that PD patients who are treated with L-dopa displayed several features including phase advance of plasma melatonin levels and delayed sleep onset [49–52]. Zheng found that melatonin could alleviate the behavioral deficits in APP695 transgenic mouse model [53]. Reduced amplitude and timing of peak levels have been also reported in AD patients [54]. However, Clifford found that melatonin cannot be an effective soporific agent in people with Alzheimer's disease [55]. In patients with HD, the timing of the evening peak in melatonin level was delayed but the total secretion volume did not change [56].

The increased oxidative stress is another important feature in the three above-mentioned neurodegenerative diseases. Increased oxidative damage may result in the death of neurons, astrocytes, and oligodendrocytes and even the brain structures. The mouse model evidence between circadian clock and reactive oxygen species (ROS) was reported by Kondratov, and he found that the mice deficient with *Bmal1* had premature aging and significantly increased ROS level in several tissues [57]. Another promising data that links the circadian clock regulation and redox status was observed in human red blood cells, and it showed that antioxidant proteins peroxiredoxins in these cells undergo 24-h cycles [5]. In Parkinson's disease, Lewy bodies contain the protein oxidation and lipid peroxidation products. In Alzheimer's disease, amyloid β plaques and neurofibrillary tangles constitute oxidative damage to proteins and lipids. Besides, both the PD and HD clinical evidence implicated the impairment of mitochondrial function which is related to ROS level.

Autophagy is the natural, regulated, destructive mechanism of the cell that disassembles unnecessary or dysfunctional components [58]. However, the physiological significance of these autophagy rhythms remains unclear. The rhythmic expression of autophagy or autophagy-related genes/proteins has been reported in mouse (liver, skeletal muscle), rat (kidney and retina), and zebrafish (larva and peripheral tissue) [59–62]. Recently, Pan's lab found that the autophagy marker proteins LC3 and Beclin1 showed rhythmic expression in hippocampus of the mouse brain, and the sleep fragmentation disrupted the rhythmicity of these two proteins [63]. These evidences implicated the link between circadian clock and autophagy degradation pathway, and the cycling of the autophagy activity may be involved in the neuroprotection roles to clear the accumulated and aggregated proteins and damaged mitochondria, which contribute to the development of the neurodegenerative disease and other age-related brain dysfunctions.

9.3   The Manifestation of Circadian Disruption in PD

9.3.1   Sleep-Wake Cycle

Sleep-wake cycle is the most typical circadian rhythm, and it affects about 60–70% of PD patients, only next to psychiatric-related problems. It includes insomnia (difficulty in falling or staying asleep), excessive daytime somnolence, rapid eye movement sleep disorder (RBD), and so on. Lack of sleep time and poor sleep quality not only have a bad impact on the quality of life for patients but also may accelerate the development of the primary disease itself [64].

Insomnia was the most common sleep disorder in the patients with PD. Studies using non-motor symptoms questionnaire showed that about half of PD patients complained about insomnia. There are several factors responsible for the insomnia, such as dopaminergic signaling, brainstem accumulation of α-synuclein, and hypocretin signaling [65]. In addition, the overnight emergence of motor symptoms like tremor, rigidity, and dyskinesia may result in awakenings and prevents from falling back to sleep again. What is more, mood disorders were important factors influencing insomnia in general population. It is reported that depression is also correlated with nocturnal sleep scores. And the severity of depression may affect sleep initiation and maintenance difficulties.

Excessive daily sleep (EDS) was more frequent in the PD patients compared to the age-matched controls. Reports showed that the prevalence ranges from 20 to 50%. And the underlying mechanism is multifactorial [66]. First, EDS may involve alterations in pathophysiological mechanisms involved in the regulation of sleep and wakefulness. Second, EDS may be linked to poor nocturnal sleep. Third, EDS could be an adverse outcome of dopaminergic therapy. Dopaminergic agonist may result in the EDS. Moreover, dopamine has the effect of promoting awakening. Taking the dopamine later in the day may influence the sleep in the night owing to the higher dopamine concentration in plasma. And thus, decreased dopamine in the daytime may result in excessive daily sleep.

RBD is a hot spot in the area of PD research. It is known as a parasomnia characterized by loss of the atonia that normally occurs during REM sleep, associated with dream enactment behavior. Of the PD patients, 40% suffered from RBD. It only can be thought as a predictor of neurodegeneration disease and can also be associated with more severe motor and non-motor manifestations.

In sum, sleep-wake cycle, which is the most important circadian rhythm, is disrupted in PD patients. It was not only affected by the non-motor syndromes and motor syndromes, but also it may correlate with the circadian system. Because increasing researches have revealed that the amount and secretion pattern of melatonin, cortisol, and hypocretin were changed, these hormones are all closely related to sleep.

9.3.2 Motor Symptoms Fluctuation in PD

As Van Hilten JJ et al. have reported, motor activity manifested as a circadian rhythm and reached the lowest level between 0:30 am and 8:00 am and the highest level at the late morning [67]. But this rhythm was modified in PD patients. In advanced PD patients, there was no obvious peak and low during the whole day. However, in mild and moderate stage patients, the pattern of motor activity was still maintained, despite the low mean level motor activity. Parkinson's disease patients often complain that the motor response to levodopa tends to be less later in the day than in the morning, and some even have no response in the evening at all. But the underlying mechanism still remains very inclusive.

The motor response to levodopa was evaluated in PD patients by Bonuccelli [68]. These patients were divided into three groups: de novo, stable, and wearing off. All these patients were given standard doses of levodopa/carbidopa, and motor activity was accessed with tapping test, walking time, and tremor score per hour. At the same time, the concentration of levodopa and 3-O-methyldopa was measured. They found that there was progressive daytime worsening in motor score with the stable and wearing-off patients, but there was no any change in de novo patients. And there was no significant difference in levodopa pharmacokinetic of each group. Thus, it is concluded that this progressive daytime attenuation of response may not relate to the pharmacokinetics of levodopa. But the duration of PD is an important risk factor for the motor fluctuation.

Motor performance also is evaluated in patients receiving different drugs [69]. These patients were divided into levodopa/carbidopa group, bromocriptine group, and "de novo" group. And a progressive daytime worsening is observed only in the group treated with levodopa, which correlated with progressive increase in 3-O-methyldopa plasma level. But this phenomenon was not found in the bromocriptine and "de novo" group. These data suggested that pharmacokinetic or pharmacodynamic factors play an important role in the weaker response to levodopa.

In sum, compared to the de novo patients, the patients in the late stage suffered more often from the motor response, which suggested that the stage of the disease was a main cause. In addition, the pharmacokinetic or pharmacodynamic factors also can affect the motor fluctuation. Whether this phenomenon was related to the circadian rhythm in the brain is to be explored.

9.3.3 Endocrine Changes in PD

Melatonin

Melatonin (N-acetyl-5-methoxytryptamine) is a natural hormone mainly produced in the mammalian pineal gland during the dark phase. It was synthesized in SCN, which receives the information from zeitgeber and relays these photoperiodic information and delivers it to the pineal gland by neuronal pathways. Melatonin can modulate the sleep pattern and shorten the time lag. These facts are suggestive that melatonin is close with sleep. In PD, two-thirds of people suffer from sleep disturbance. So several studies had been done to observe the melatonin amount in plasma.

In 1993, E. Fertl and colleague found that the amount of melatonin did not change, but the phase of melatonin expression was advanced compared the age-matched control [70]. And then, they continued to check whether the advanced phase is associated with dopamine treatment. Their results showed that the advance phase in Parkinson's disease patients is due to a central nervous dopaminergic effect, but not the disease itself. Actually, there are β-adrenergic receptors in pinealocyte, by which administration of L-dopa may lead to an increased melatonin secretion. In 2003, Bordet R. investigated the circadian melatonin pattern at different stages of PD [49]. They found administration of dopamine could render a phase advance in plasma melatonin. The amount of melatonin during the daytime was increased in PD with levodopa-related motor complications. However, in these patients, the amount of melatonin during the nighttime is decreased significantly. Until recently, some investigators paid attention to melatonin secretion pattern in PD again. They found dopaminergic treatment may increase the secretion of melatonin and induced a delayed sleep onset relative to the melatonin secretion onset [50]. That is to say, dopamine may result in uncoupling between the melatonin synthesis and sleep-wake circle, which may account for part of sleep disturbances in L-dopa-treated PD patients. In 2014, David P. Breen reported that about half of newly diagnosed patients had sleep complaints and poorer quality of life compared to those without sleep disturbances [52]. In these patients, there was a sustained reduced circulation melatonin levels during the daytime and nighttime. At the same time, Aleksandar Videnovic found the similar phenomenon in L-dopa-treated patients. They reported that the amplitude of the melatonin rhythm and 24-h AUC of circulating melatonin levels were significantly lower in PD patients [51]. There were several evidences supporting the decrease of melatonin in PD. Firstly, Lewy body formation was observed in every hypothalamus in PD patients. The lateral and posterior hypothalamic nuclei and tuberomammillary nucleus were most frequently involved and with the highest average Lewy body counts. An in vivo C-raclopride PET study was to observe the difference of dopamine D2 receptor availability between Parkinson's disease and healthy age-matched control [71]. They reported there was a reduction in hypothalamic D2 receptor availability seen in patients of PD, and this reduction may contribute to the development of endocrine, autonomic disorders, and sleep. Whether this dysfunction is related to this neurodegeneration disease itself or chronic exposure to levodopa or both is still confusing. Secondly, the following study provided information that the level of dopamine content in hypothalamic was down [72]. And postmortem also showed dopamine concentrations in the hypothalamus decreased in PD, which may partially lead to less melatonin secretion.

Cortisol
Cortisol secretion is controlled by hypothalamic-pituitary-adrenal (HPA) axis, which receives the circadian flow from the hypothalamic paraventricular nucleus. Thus, the level of cortisol in human body also displays rhythm, with the peak in the morning and the low in the night. In several studies, the cortisol secretion is disrupted in neurodegenerative disease, including PD.

A 24-h cortisol release profiles were measured, and the results showed that patients with PD secrete significantly more cortisol than the age-matched control [73]. Furthermore, levodopa therapy may not attribute to this endocrine abnormality, with dopamine not playing a major role in modulating of HPA system activity. In early Parkinson's disease, this hypercortisol also can be found. Similarly, in MPTP-treated dogs, the experiments also showed a significant increase in 24-h plasma cortisol concentration [74].

This may be due to the dysfunction of SCN, because the image and neuropathological study suggested SCN was affected in the PD patients.

9.3.4   Core Body Temperature

Body temperature is considered to be a marker rhythm of circadian system. Several studies reported that the body temperature was lower in PD than in the healthy controls. The mesor of the core body temperature was significantly lower in the PD patients [75]. In addition, the difference between the mesor and nadir temperature was reduced, which was related to the severity of RBD. The different effect of naloxone on the body temperature was observed in the postmenopausal women with and without PD [76]. In the postmenopausal women without PD, the body temperature was reduced by the naloxone infusion. But in the postmenopausal women with PD, this pattern was modified, suggesting that there was an impaired thermoregulation by endogenous opioid system with PD. At the same time, this study also provided evidence that the body temperature with PD was lower than that of controls.

Like PD, in other neurodegeneration disease, the dysfunction of thermoregulation was also observed. It is found that the physiological nocturnal fall of the body core temperature is blunted in MSA patients, but not in PD. And Suzuki K et al. reported that the circadian rhythmicity of the core body temperature was changed and the amplitude was also lower in patients with progressive supranuclear palsy than those with PD [77].

9.3.5   Ambulatory Blood Pressure and Heart Rate

The changes of heart rate and blood pressure have a 24-h period, based on the endogenous circadian rhythms and sleep-activity rhythm. Usually, the mean arterial pressure decreased by 10–20% in the sleep, compared to that in the awake time. But in most patients with PD, this nocturnal fall of blood pressure was lost independent of orthostatic hypotension [78]. A study by Ejaz AA identified that the circadian rhythm was reversed in 93% patients with Parkinson's disease using the method of 24-h ambulatory blood pressure monitoring (ABPM) [79]. At the same time, they also found all these patients had postprandial hypotension and nocturnal hypertension. And other study also found there was abnormal physiological circadian rhythm in 71.1% of patients, which was higher than that in the controls. Moreover, impaired blood pressure pattern correlated with the prevalence of autonomic disorders [80].

9.3.6   Visual Contrastibility

Retina can be thought of as an endogenous circadian clock, which could receive the sunshine signal. To adapt to the environment around, the retina can modulate kinds of circadian rhythms, such as rode-cone balance, visual sensitivity and electroretinogram (ERG) b-wave amplitude, and dopamine synthesis. In mammals, retinal clock and its outputs may influence the trophic situation in eyes. So, the dysfunction of retinal clock may result in the less susceptibility of photoreceptor and the low survival rate in retinal degeneration of animal models. Dopamine, as the major catecholamine in the retina of vertebrate, plays an important role in the adaption to the light. Using a PERIOD2:LUCIFERASE fusion protein knock-in model, researchers reported that the dopamine can regulate the PER proteins and play a significant role in resetting the retinal circadian clock by D1 receptor [81].

In the PD patients, the contrast sensitivity fluctuates during the whole day [82]. The contrast sensitivity was tested in PD patients and controls, at 2-h intervals from 8:30 AM. At the beginning of the day, there was no any difference between these two groups. But the sensitivity of PD was significantly worse in the late day than that of control. In other words, the contrast sensitivity was kept unchanged during the test in the control group. However, the PD patients were worse at 2:30 PM than at 8:30 PM.

9.3.7 Clock Gene

It is well known that clock gene is an important part in biorhythm. And increasing evidence showed that these clock genes are not only expressed in SCN but also in many peripheral organs. There are several circadian genes called as key clock genes, such as Clock, Bmal, Period (Per1, Per2, Per3), and cryptochrome (Cry1, Cry2). In recent years, several researches have aimed to study the desynchronized oscillatory in PD patients at the molecular level.

In the whole blood, the mRNA expression of Bmal1 was significantly lower in PD patients [83, 84]. And the Bmal1 mRNA level had a correlation with motor severity and sleep quality. Recently, genetic polymorphisms were tested in PD patients. It was found that Bmal1 expression declined in PD, and it was reported that the Bmal1 variant was related to the occurrence of the tremor dominant subtype, while the Per1 variant was to the postural instability and gait difficulty dominant subtype [85]. This suggests that abnormal expression of the clock genes may not only affect the biorhythm but also plays an etiological role in PD. Modifications of these clock genes have also been studied in PD animal models. A reduction of the daily striatal Per2 expression has been reported in 6-OHDA lesioned rat model [86]. In line with this, the daily pulse of Per2, Cry1, and Bmal1 was decreased in SCN induced by rotenone, which could be partially restored by melatonin administration [87].

However, what is the mechanism underlying the clock gene alterations in PD? As we know, DNA methylation plays an important role in downregulating the expression of clock genes, which aims to the CpG island. Abnormal CpG methylation has been found in neurodegenerative disorders, including PD. In PD, the methylation of the NPAS2 (the paralog of clock) promoter is decreased and thus the NPAS2 expression is increased. This, in turn, would promote the expression of Rora and Rev-erba, which serve as the main regulators of Bmal1 expression. Above all, the epigenetic alterations in NPSA2 expression may contribute to the alteration of Bmal1 and Bmal2 in the leukocytes of PD patients [88].

In addition, the regular expression of clock genes is closely related to other circadian functions. Therefore, the non-motor symptoms and the fluctuation of motor ability may be partially relevant with abnormal clock genes. Several studies show that the molecular regulators of the circadian clock have different influences on sleep patterns. For example, in NPAS2-deficient mice, non-REM sleep is reduced and more sleep time is required following sleep deprivation [89]. Researchers also demonstrated that locomotor activity in both constant darkness and light–dark cycles was impaired and total activity levels were reduced in Bmal1 knockout mice [90]. In addition, deletion of Bmal1 also induced the attenuated rhythm of the body temperature [91]. Some clock genes also modulate the visual information processing to light. Also, melatonin could induce phase advance of Per1 and Per2 at dusk through activation of protein kinase C [92].

In sum, PD is related to the disrupted expression profile of clock genes, which may interact with other physiological activities.

9.4 The Application of Recovering the Circadian Rhythm in the Therapy of PD

Circadian therapy is aimed at regulating the biorhythm by changing the external zeitgebers. And the main external zeitgebers contain physical activity and dietary and social schedules. Increasing evidences suggest that circadian oscillation is disturbed in PD and that circadian regulation may become a new target for therapeutic intervention.

Light therapy has previously been used for seasonal affective disorder. Many promising initiations have extended this therapy to a wider disease spectrum. Dopamine is a chemical messenger for light adaptation, and light exposure could increase dopamine activity in retina. Recently, circadian light entrainment is increasingly applied to PD patients attempting to reset the biologic clock. Paus et al. found that half an hour bright light therapy (BLT) in the morning could result in a marked improvement in motor ability and depression [93]. In another study, PD patients were arranged to BLT once daily at an intensity of 1000–1500 lx, before the usual time of sleep onset for 1 h [94]. Two weeks later, bradykinesia and rigidity was ameliorated in most patients, as well as elevated mood, improved sleep, and reduced demands for drugs. Later, this group carried out a retrospective and open-label study in 129 PD patients to assess the systematic application of BLT [95]. The results confirmed that BLT is good to the motor performance recovery of PD patients and decreased the dosage of dopamine replacement drugs. Meanwhile, BLT could improve the mood and sleep quality in PD patient. So, BLT has a positive effect on motor and non-motor manifestations in PD. Moreover, due to the advantages of non-invasion, low cost, and convenient usage, BLT is a good option worth considering. However, more research should be focused on early diagnosed or more severe PD patients to delay the progress or relieve the burden.

On the other hand, physical exercise could delay disease progression in some way, which may be explained as it could regulate the biorhythm. Yamanaka et al.

think that physical exercise could enhance the sleep-wake cycle, which was independent of the circadian pacemaker [96]. Morning and evening exercise differentially regulate the autonomic nervous system [97]. More evidence was provided in animal models. For example, voluntary wheel-running could affect physiological circadian rhythms and delay the phase of peripheral Per2 expression [98]. Also, physical exercises including Argentine Tango and Tai Ji Quan may be appropriate choices for postural instability therapy in PD. However, few exploratory studies have examined the effect of physical exercise on circadian systems in PD patients.

9.5 The Uncovered Question in the Circadian Rhythm of PD

To date, many efforts are made to cover the relationship between circadian dysfunction and PD. But it is still unable to confirm whether circadian disruption is a causal factor for PD pathogenesis or a consequence of PD progression. On one hand, circadian disruption could exacerbate some pathological events of PD, which is quite definite. Several molecular mechanisms have been reported. For example, neuro-inflammation is controlled by the intrinsic circadian clock in the microglia, and circadian dysfunction could induce robust neuro-inflammation responses [99]. As we know, a-synuclein aggregate, the main pathology of PD, is mostly degraded via autophagy. But the expression of autophagy-related proteins in the brain displayed a 24-h rhythm, which could be blunted by sleep fragmentation [63]. Mitochondrion is targeted by most neurotoxins of PD. Its dysfunction is associated with PD-related genes, such as PINK1, DJ-1, and PARKIN. Interestingly, the function of mitochondrial respiration also displayed in a diurnal manner, and it was regulated by the clock proteins [100]. On the other hand, the circadian system may also be altered along with the PD progression. SCN is found with the a-synuclein deposit, which could induce the death of the key cells in SCN. Moreover, the circadian dysfunction is often observed in neurotoxin-induced PD models, which may hint that the abnormal biorhythm may be secondary to dopamine depletion [101]. In addition, several circadian manifestations in PD patients are much more severe in the late stage than that in the early stage, which indicates that they may be consequences of the disease or the long-term drug medication. Therefore, the disrupted biorhythm and the disease itself may interact as both cause and effect.

References

1. Edgar RS, et al. Peroxiredoxins are conserved markers of circadian rhythms. Nature. 2012;485:459–64. https://doi.org/10.1038/nature11088.
2. Konopka RJ, Benzer S. Clock mutants of *Drosophila melanogaster*. Proc Natl Acad Sci U S A. 1971;68:2112–6.
3. Lowrey PL, Takahashi JS. Mammalian circadian biology: elucidating genome-wide levels of temporal organization. Annu Rev Genomics Hum Genet. 2004;5:407–41. https://doi.org/10.1146/annurev.genom.5.061903.175925.

4. Hastings MH, Reddy AB, Maywood ES. A clockwork web: circadian timing in brain and periphery, in health and disease. Nat Rev Neurosci. 2003;4:649–61. https://doi.org/10.1038/nrn1177.

5. O'Neill JS, Reddy AB. Circadian clocks in human red blood cells. Nature. 2011;469:498–503. https://doi.org/10.1038/nature09702.

6. Buhr ED, Takahashi JS. Molecular components of the mammalian circadian clock. Handb Exp Pharmacol. 2013;217:3–27. https://doi.org/10.1007/978-3-642-25950-0_1.

7. Dibner C, Schibler U, Albrecht U. The mammalian circadian timing system: organization and coordination of central and peripheral clocks. Annu Rev Physiol. 2010;72:517–49. https://doi.org/10.1146/annurev-physiol-021909-135821.

8. Liu AC, et al. Intercellular coupling confers robustness against mutations in the SCN circadian clock network. Cell. 2007;129:605–16. https://doi.org/10.1016/j.cell.2007.02.047.

9. Mohawk JA, Takahashi JS. Cell autonomy and synchrony of suprachiasmatic nucleus circadian oscillators. Trends Neurosci. 2011;34:349–58. https://doi.org/10.1016/j.tins.2011.05.003.

10. Buhr ED, Yoo SH, Takahashi JS. Temperature as a universal resetting cue for mammalian circadian oscillators. Science. 2010;330:379–85. https://doi.org/10.1126/science.1195262.

11. Yang X, Lamia KA, Evans RM. Nuclear receptors, metabolism, and the circadian clock. Cold Spring Harb Symp Quant Biol. 2007;72:387–94. https://doi.org/10.1101/sqb.2007.72.058.

12. Huang N, et al. Crystal structure of the heterodimeric CLOCK:BMAL1 transcriptional activator complex. Science. 2012;337:189–94. https://doi.org/10.1126/science.1222804.

13. Kume K, et al. mCRY1 and mCRY2 are essential components of the negative limb of the circadian clock feedback loop. Cell. 1999;98:193–205.

14. Busino L, et al. SCFFbxl3 controls the oscillation of the circadian clock by directing the degradation of cryptochrome proteins. Science. 2007;316:900–4. https://doi.org/10.1126/science.1141194.

15. Siepka SM, et al. Circadian mutant overtime reveals F-box protein FBXL3 regulation of cryptochrome and period gene expression. Cell. 2007;129:1011–23. https://doi.org/10.1016/j.cell.2007.04.030.

16. Sato TK, et al. A functional genomics strategy reveals Rora as a component of the mammalian circadian clock. Neuron. 2004;43:527–37. https://doi.org/10.1016/j.neuron.2004.07.018.

17. Preitner N, et al. The orphan nuclear receptor REV-ERBalpha controls circadian transcription within the positive limb of the mammalian circadian oscillator. Cell. 2002;110:251–60.

18. Ukai-Tadenuma M, et al. Delay in feedback repression by cryptochrome 1 is required for circadian clock function. Cell. 2011;144:268–81. https://doi.org/10.1016/j.cell.2010.12.019.

19. Liu AC, et al. Redundant function of REV-ERBalpha and beta and non-essential role for Bmal1 cycling in transcriptional regulation of intracellular circadian rhythms. PLoS Genet. 2008;4:e1000023. https://doi.org/10.1371/journal.pgen.1000023.

20. Crosio C, Cermakian N, Allis CD, Sassone-Corsi P. Light induces chromatin modification in cells of the mammalian circadian clock. Nat Neurosci. 2000;3:1241–7. https://doi.org/10.1038/81767.

21. Doi M, Hirayama J, Sassone-Corsi P. Circadian regulator CLOCK is a histone acetyltransferase. Cell. 2006;125:497–508. https://doi.org/10.1016/j.cell.2006.03.033.

22. Duong HA, Robles MS, Knutti D, Weitz CJ. A molecular mechanism for circadian clock negative feedback. Science. 2011;332:1436–9. https://doi.org/10.1126/science.1196766.

23. Naruse Y, et al. Circadian and light-induced transcription of clock gene Per1 depends on histone acetylation and deacetylation. Mol Cell Biol. 2004;24:6278–87. https://doi.org/10.1128/MCB.24.14.6278-6287.2004.

24. Feng D, et al. A circadian rhythm orchestrated by histone deacetylase 3 controls hepatic lipid metabolism. Science. 2011;331:1315–9. https://doi.org/10.1126/science.1198125.

25. Keene AC, Duboue ER. The origins and evolution of sleep. J Exp Biol. 2018;221:jeb159533. https://doi.org/10.1242/jeb.159533.

26. Spiegel K, Leproult R, Van Cauter E. Impact of sleep debt on metabolic and endocrine function. Lancet. 1999;354:1435–9. https://doi.org/10.1016/S0140-6736(99)01376-8.

27. Ding F, et al. Changes in the composition of brain interstitial ions control the sleep-wake cycle. Science. 2016;352:550–5. https://doi.org/10.1126/science.aad4821.
28. Tononi G, Cirelli C. Sleep and the price of plasticity: from synaptic and cellular homeostasis to memory consolidation and integration. Neuron. 2014;81:12–34. https://doi.org/10.1016/j.neuron.2013.12.025.
29. de Vivo L, et al. Ultrastructural evidence for synaptic scaling across the wake/sleep cycle. Science. 2017;355:507–10. https://doi.org/10.1126/science.aah5982.
30. Marshall L, Helgadottir H, Molle M, Born J. Boosting slow oscillations during sleep potentiates memory. Nature. 2006;444:610–3. https://doi.org/10.1038/nature05278.
31. Blum ID, Bell B, Wu MN. Time for bed: genetic mechanisms mediating the circadian regulation of sleep. Trends Genet. 2018;34:379–88. https://doi.org/10.1016/j.tig.2018.01.001.
32. Toh KL, et al. An hPer2 phosphorylation site mutation in familial advanced sleep phase syndrome. Science. 2001;291:1040–3.
33. Patke A, et al. Mutation of the human circadian clock gene CRY1 in familial delayed sleep phase disorder. Cell. 2017;169:203–215 e213. https://doi.org/10.1016/j.cell.2017.03.027.
34. Zhang L, et al. A PERIOD3 variant causes a circadian phenotype and is associated with a seasonal mood trait. Proc Natl Acad Sci U S A. 2016;113:E1536–44. https://doi.org/10.1073/pnas.1600039113.
35. Xu Y, et al. Functional consequences of a CKIdelta mutation causing familial advanced sleep phase syndrome. Nature. 2005;434:640–4. https://doi.org/10.1038/nature03453.
36. Katzenberg D, et al. A CLOCK polymorphism associated with human diurnal preference. Sleep. 1998;21:569–76.
37. Hirano A, et al. A Cryptochrome 2 mutation yields advanced sleep phase in humans. elife. 2016;5:e16695. https://doi.org/10.7554/eLife.16695.
38. He Y, et al. The transcriptional repressor DEC2 regulates sleep length in mammals. Science. 2009;325:866–70. https://doi.org/10.1126/science.1174443.
39. Pellegrino R, et al. A novel BHLHE41 variant is associated with short sleep and resistance to sleep deprivation in humans. Sleep. 2014;37:1327–36. https://doi.org/10.5665/sleep.3924.
40. Ebisawa T, et al. Association of structural polymorphisms in the human period3 gene with delayed sleep phase syndrome. EMBO Rep. 2001;2:342–6. https://doi.org/10.1093/embo-reports/kve070.
41. Wulff K, Gatti S, Wettstein JG, Foster RG. Sleep and circadian rhythm disruption in psychiatric and neurodegenerative disease. Nat Rev Neurosci. 2010;11:589–99. https://doi.org/10.1038/nrn2868.
42. Sterniczuk R, Dyck RH, Laferla FM, Antle MC. Characterization of the 3xTg-AD mouse model of Alzheimer's disease: part 1. Circadian changes. Brain Res. 2010;1348:139–48. https://doi.org/10.1016/j.brainres.2010.05.013.
43. Kudo T, Loh DH, Truong D, Wu Y, Colwell CS. Circadian dysfunction in a mouse model of Parkinson's disease. Exp Neurol. 2011;232:66–75. https://doi.org/10.1016/j.expneurol.2011.08.003.
44. Oakeshott S, et al. Circadian abnormalities in motor activity in a BAC transgenic mouse model of Huntington's disease. PLoS Curr. 2011;3:RRN1225. https://doi.org/10.1371/currents.RRN1225.
45. Kondratova AA, Kondratov RV. The circadian clock and pathology of the ageing brain. Nat Rev Neurosci. 2012;13:325–35. https://doi.org/10.1038/nrn3208.
46. Altun A, Ugur-Altun B. Melatonin: therapeutic and clinical utilization. Int J Clin Pract. 2007;61:835–45. https://doi.org/10.1111/j.1742-1241.2006.01191.x.
47. Reiter RJ, Acuna-Castroviejo D, Tan DX, Burkhardt S. Free radical-mediated molecular damage. Mechanisms for the protective actions of melatonin in the central nervous system. Ann N Y Acad Sci. 2001;939:200–15.
48. Hardeland R. Antioxidative protection by melatonin: multiplicity of mechanisms from radical detoxification to radical avoidance. Endocrine. 2005;27:119–30.
49. Bordet R, et al. Study of circadian melatonin secretion pattern at different stages of Parkinson's disease. Clin Neuropharmacol. 2003;26:65–72.

50. Bolitho SJ, et al. Disturbances in melatonin secretion and circadian sleep-wake regulation in Parkinson disease. Sleep Med. 2014;15:342–7. https://doi.org/10.1016/j.sleep.2013.10.016.
51. Videnovic A, et al. Circadian melatonin rhythm and excessive daytime sleepiness in Parkinson disease. JAMA Neurol. 2014;71:463–9. https://doi.org/10.1001/jamaneurol.2013.6239.
52. Breen DP, et al. Sleep and circadian rhythm regulation in early Parkinson disease. JAMA Neurol. 2014;71:589–95. https://doi.org/10.1001/jamaneurol.2014.65.
53. Feng Z, et al. Melatonin alleviates behavioral deficits associated with apoptosis and cholinergic system dysfunction in the APP 695 transgenic mouse model of Alzheimer's disease. J Pineal Res. 2004;37:129–36. https://doi.org/10.1111/j.1600-079X.2004.00144.x.
54. Mishima K, et al. Melatonin secretion rhythm disorders in patients with senile dementia of Alzheimer's type with disturbed sleep-waking. Biol Psychiatry. 1999;45:417–21.
55. Singer C, et al. A multicenter, placebo-controlled trial of melatonin for sleep disturbance in Alzheimer's disease. Sleep. 2003;26:893–901.
56. Aziz NA, et al. Delayed onset of the diurnal melatonin rise in patients with Huntington's disease. J Neurol. 2009;256:1961–5. https://doi.org/10.1007/s00415-009-5196-1.
57. Kondratov RV, Vykhovanets O, Kondratova AA, Antoch MP. Antioxidant N-acetyl-L-cysteine ameliorates symptoms of premature aging associated with the deficiency of the circadian protein BMAL1. Aging (Albany NY). 2009;1:979–87. https://doi.org/10.18632/aging.100113.
58. Klionsky DJ. Autophagy revisited: a conversation with Christian de Duve. Autophagy. 2008;4:740–3.
59. Ma D, Panda S, Lin JD. Temporal orchestration of circadian autophagy rhythm by C/EBPbeta. EMBO J. 2011;30:4642–51. https://doi.org/10.1038/emboj.2011.322.
60. Reme C, Wirz-Justice A, Rhyner A, Hofmann S. Circadian rhythm in the light response of rat retinal disk-shedding and autophagy. Brain Res. 1986;369:356–60.
61. Pfeifer U, Scheller H. A morphometric study of cellular autophagy including diurnal variations in kidney tubules of normal rats. J Cell Biol. 1975;64:608–21.
62. Huang G, Zhang F, Ye Q, Wang H. The circadian clock regulates autophagy directly through the nuclear hormone receptor Nr1d1/rev-erbalpha and indirectly via Cebpb/(C/ebpbeta) in zebrafish. Autophagy. 2016;12:1292–309. https://doi.org/10.1080/15548627.2016.1183843.
63. He Y, et al. Circadian rhythm of autophagy proteins in hippocampus is blunted by sleep fragmentation. Chronobiol Int. 2016;33:553–60. https://doi.org/10.3109/07420528.2015.1137581.
64. Li S, Wang Y, Wang F, Hu LF, Liu CF. A new perspective for Parkinson's disease: circadian rhythm. Neurosci Bull. 2017;33:62–72. https://doi.org/10.1007/s12264-016-0089-7.
65. Rothman SM, Mattson MP. Sleep disturbances in Alzheimer's and Parkinson's diseases. NeuroMolecular Med. 2012;14:194–204. https://doi.org/10.1007/s12017-012-8181-2.
66. Shen Y, Huang JY, Li J, Liu CF. Excessive daytime sleepiness in Parkinson's disease: clinical implications and management. Chin Med J. 2018;131:974–81. https://doi.org/10.4103/0366-6999.229889.
67. van Hilten JJ, et al. Diurnal effects of motor activity and fatigue in Parkinson's disease. J Neurol Neurosurg Psychiatry. 1993;56:874–7.
68. Bonuccelli U, et al. Diurnal motor variations to repeated doses of levodopa in Parkinson's disease. Clin Neuropharmacol. 2000;23:28–33.
69. Piccini P, et al. Diurnal worsening in Parkinson patients treated with levodopa. Riv Neurol. 1991;61:219–24.
70. Fertl E, Auff E, Doppelbauer A, Waldhauser F. Circadian secretion pattern of melatonin in Parkinson's disease. J Neural Transm Park Dis Dement Sect. 1991;3:41–7.
71. Breen DP, et al. Hypothalamic volume loss is associated with reduced melatonin output in Parkinson's disease. Mov Disord. 2016;31:1062–6. https://doi.org/10.1002/mds.26592.
72. Bogaerts V, Theuns J, van Broeckhoven C. Genetic findings in Parkinson's disease and translation into treatment: a leading role for mitochondria? Genes Brain Behav. 2008;7:129–51. https://doi.org/10.1111/j.1601-183X.2007.00342.x.

73. Hartmann A, Veldhuis JD, Deuschle M, Standhardt H, Heuser I. Twenty-four hour cortisol release profiles in patients with Alzheimer's and Parkinson's disease compared to normal controls: ultradian secretory pulsatility and diurnal variation. Neurobiol Aging. 1997;18:285–9.

74. Mizobuchi M, Hineno T, Kakimoto Y, Hiratani K. Increase of plasma adrenocorticotrophin and cortisol in 1-methyl-4-phenyl-1,2,3,6-tetrahydropyridine (MPTP)-treated dogs. Brain Res. 1993;612:319–21.

75. Zhong G, Bolitho S, Grunstein R, Naismith SL, Lewis SJ. The relationship between thermoregulation and REM sleep behaviour disorder in Parkinson's disease. PLoS One. 2013;8:e72661. https://doi.org/10.1371/journal.pone.0072661.

76. Cagnacci A, et al. Effect of naloxone on body temperature in postmenopausal women with Parkinson's disease. Life Sci. 1990;46:1241–7.

77. Suzuki K, et al. Circadian variation of core body temperature in Parkinson disease patients with depression: a potential biological marker for depression in Parkinson disease. Neuropsychobiology. 2007;56:172–9. https://doi.org/10.1159/000119735.

78. Schmidt C, et al. Loss of nocturnal blood pressure fall in various extrapyramidal syndromes. Mov Disord. 2009;24:2136–42. https://doi.org/10.1002/mds.22767.

79. Ejaz AA, Sekhon IS, Munjal S. Characteristic findings on 24-h ambulatory blood pressure monitoring in a series of patients with Parkinson's disease. Eur J Intern Med. 2006;17:417–20. https://doi.org/10.1016/j.ejim.2006.02.020.

80. Berganzo K, et al. Nocturnal hypertension and dysautonomia in patients with Parkinson's disease: are they related? J Neurol. 2013;260:1752–6. https://doi.org/10.1007/s00415-013-6859-5.

81. Ruan GX, Allen GC, Yamazaki S, McMahon DG. An autonomous circadian clock in the inner mouse retina regulated by dopamine and GABA. PLoS Biol. 2008;6:e249. https://doi.org/10.1371/journal.pbio.0060249.

82. Struck LK, Rodnitzky RL, Dobson JK. Circadian fluctuations of contrast sensitivity in Parkinson's disease. Neurology. 1990;40:467–70.

83. Cai Y, Liu S, Sothern RB, Xu S, Chan P. Expression of clock genes Per1 and Bmal1 in total leukocytes in health and Parkinson's disease. Eur J Neurol. 2010;17:550–4. https://doi.org/10.1111/j.1468-1331.2009.02848.x.

84. Ding H, et al. Decreased expression of Bmal2 in patients with Parkinson's disease. Neurosci Lett. 2011;499:186–8. https://doi.org/10.1016/j.neulet.2011.05.058.

85. Gu Z, et al. Association of ARNTL and PER1 genes with Parkinson's disease: a case-control study of Han Chinese. Sci Rep. 2015;5:15891. https://doi.org/10.1038/srep15891.

86. Hood S, et al. Endogenous dopamine regulates the rhythm of expression of the clock protein PER2 in the rat dorsal striatum via daily activation of D2 dopamine receptors. J Neurosci. 2010;30:14046–58. https://doi.org/10.1523/JNEUROSCI.2128-10.2010.

87. Mattam U, Jagota A. Daily rhythms of serotonin metabolism and the expression of clock genes in suprachiasmatic nucleus of rotenone-induced Parkinson's disease male Wistar rat model and effect of melatonin administration. Biogerontology. 2015;16:109–23. https://doi.org/10.1007/s10522-014-9541-0.

88. Lin Q, et al. Promoter methylation analysis of seven clock genes in Parkinson's disease. Neurosci Lett. 2012;507:147–50. https://doi.org/10.1016/j.neulet.2011.12.007.

89. Dudley CA, et al. Altered patterns of sleep and behavioral adaptability in NPAS2-deficient mice. Science. 2003;301:379–83. https://doi.org/10.1126/science.1082795.

90. Bunger MK, et al. Mop3 is an essential component of the master circadian pacemaker in mammals. Cell. 2000;103:1009–17.

91. Xie Z, et al. Smooth-muscle BMAL1 participates in blood pressure circadian rhythm regulation. J Clin Invest. 2015;125:324–36. https://doi.org/10.1172/JCI76881.

92. Kandalepas PC, Mitchell JW, Gillette MU. Melatonin signal transduction pathways require E-box-mediated transcription of Per1 and Per2 to reset the SCN clock at dusk. PLoS One. 2016;11:e0157824. https://doi.org/10.1371/journal.pone.0157824.

93. Paus S, et al. Bright light therapy in Parkinson's disease: a pilot study. Mov Disord. 2007;22:1495–8. https://doi.org/10.1002/mds.21542.

94. Willis GL, Turner EJ. Primary and secondary features of Parkinson's disease improve with strategic exposure to bright light: a case series study. Chronobiol Int. 2007;24:521–37. https://doi.org/10.1080/07420520701420717.

95. Willis GL, Moore C, Armstrong SM. A historical justification for and retrospective analysis of the systematic application of light therapy in Parkinson's disease. Rev Neurosci. 2012;23:199–226. https://doi.org/10.1515/revneuro-2011-0072.

96. Yamanaka Y, et al. Differential regulation of circadian melatonin rhythm and sleep-wake cycle by bright lights and nonphotic time cues in humans. Am J Physiol Regul Integr Comp Physiol. 2014;307:R546–57. https://doi.org/10.1152/ajpregu.00087.2014.

97. Yamanaka Y, et al. Morning and evening physical exercise differentially regulate the autonomic nervous system during nocturnal sleep in humans. Am J Physiol Regul Integr Comp Physiol. 2015;309:R1112–21. https://doi.org/10.1152/ajpregu.00127.2015.

98. Yasumoto Y, Nakao R, Oishi K. Free access to a running-wheel advances the phase of behavioral and physiological circadian rhythms and peripheral molecular clocks in mice. PLoS One. 2015;10:e0116476. https://doi.org/10.1371/journal.pone.0116476.

99. Fonken LK, et al. Microglia inflammatory responses are controlled by an intrinsic circadian clock. Brain Behav Immun. 2015;45:171–9. https://doi.org/10.1016/j.bbi.2014.11.009.

100. Neufeld-Cohen A, et al. Circadian control of oscillations in mitochondrial rate-limiting enzymes and nutrient utilization by PERIOD proteins. Proc Natl Acad Sci U S A. 2016;113:E1673–82. https://doi.org/10.1073/pnas.1519650113.

101. Li SY, et al. Long-term levodopa treatment accelerates the circadian rhythm dysfunction in a 6-hydroxydopamine rat model of Parkinson's disease. Chin Med J. 2017;130:1085–92. https://doi.org/10.4103/0366-6999.204920.

Conclusive Remarks

10

Yun Shen and Chun-Feng Liu

Abstract

Sleep disorders are one of the most common non-motor manifestations in Parkinson's disease (PD). We make a brief summary of the following disorders of sleep and wakefulness in PD: insomnia, REM sleep behavior disorder, excessive daytime sleepiness, restless legs syndrome, sleep disordered breathing, and circadian rhythm disorders. Additionally, we suggest some fields of further research focusing on sleep and PD.

Keywords

Sleep · Sleep disorders · Parkinson's disease

Parkinson's disease (PD) is the second most common neurodegenerative disorder in the world. In China, the prevalence of PD has been estimated to be 30–87%, and approximately 48–89% of Chinese patients with PD have been shown to be affected by sleep disorders. In recent decades, there have been major advances in our understanding of the relationship between sleep disorders and PD, yet many questions remain unanswered.

Epidemiological studies have demonstrated that sleep disorders are associated with a decline in cognitive performance, productivity, mood, and quality of life, as well as with major social, medical, and economic impacts. At present, the pathophysiology of sleep–wake disturbances in PD remains largely unknown, although the etiology is most likely multifactorial. Alterations of pathophysiological

Y. Shen · C.-F. Liu (✉)
Department of Neurology, The Second Affiliated Hospital of Soochow University,
Suzhou, China
e-mail: liuchunfeng@suda.edu.cn

mechanisms are thought to underlie several processes, including sleep–wake regulatory centers, overnight emergence of motor symptoms, adverse effects of antiparkinsonian medications, psychiatric symptoms, and sleep fragmentation caused by multiple factors [1, 2]. However, recent genetic studies of PD had identified multiple genes and loci which might be associated with sleep disorders. Genetic studies of RBD offered some new insights that GBA mutations and MAPT loci were associated with RBD. Moreover, some genes and genetic loci were associated with RLS: MEIS1, BTBD9, PTPRD, MAP 2K5/SKOR1, TOX3, and RLS1-8 [3].

Sleep disorders in PD can be classified into two categories: disturbances of sleep and disturbances of wakefulness. The most common disorders include insomnia, rapid eye movement sleep behavior disorder (RBD), excessive daytime sleepiness (EDS), restless legs syndrome (RLS), sleep-disordered breathing (SDB), and disruptions of circadian rhythms.

Insomnia is defined as a persistent difficulty with sleep initiation, duration, consolidation, or quality. The patient might report some of the following symptoms related to the nighttime sleep difficulty, such as fatigue, memory impairment, daytime sleepiness, and so on. It is the most common sleep disorder in PD cases, and these patients usually report difficulties with sleep onset and sleep maintenance. Sleep fragmentation is a key indicator of sleep maintenance insomnia. Psychiatric symptoms have been shown to have a negative impact on sleep quality, particularly with respect to depression, which commonly results in early-morning awakenings. Recently, there has been an increase in the amount of research that explores cognitive behavior therapy for insomnia in patients with PD [4].

RBD is a parasomnia and associated with dream enactment behavior. These vocalizations or behaviors often correlate with dream mentation, leading to the frequent report of "acting out one's dreams." RBD has been estimated to affect 22–60% of Chinese patients with PD. While this disorder has been an area of intensive investigation over the past few decades, the mechanisms underlying it remain poorly understood. At present, diagnosis of RBD requires polysomnography.

Importantly, RBD can act as a window into long-term brain health, as it is both a symptom of early stage α-synucleinopathy and a potential marker of more severe disease manifestations in PD. Therefore, RBD will likely be a priority in future research. Moreover, based on our previous research, we hypothesize that rapid eye movement sleep without atonia is associated with the severity of PD illness and might continue to develop as PD progresses [5].

RBD are more strongly associated with neurodegenerative diseases than other sleep disorders. Idiopathic RBD can have subtle prodromal neurodegenerative abnormalities, including hyposmia, constipation, orthostatic hypotension, autonomic dysfunction, and abnormalities in gait, neuroimaging, and neurophysiological tests. In addition, 74% of patients with RBD may meet the Movement Disorders Society criteria for a diagnosis of prodromal PD [6]. Furthermore, up to 90.9% of RBD cases ultimately develop a neurodegenerative disease over the course of longitudinal follow-up. Finally, RBD is associated with more severe motor and non-motor manifestations in patients with PD than other groups. Therefore, clinicians should pay attention to the clinical course of RBD and its rate of phenoconversion and need to administer future neuroprotective therapies that can modify the course, delay the onset, or prevent the development of the disabling manifestations of PD [7].

EDS affects approximately 13–47% of Chinese patients with PD and has an annual incidence of 6%. EDS can affect both the motor and non-motor symptoms of patients with PD. At present, there are few options for the pharmacological management of EDS in PD. However, in these cases, clinicians should note the impact of dopaminergic therapy, especially any adverse effects of dopamine agonists. Future studies are needed to develop long-term therapies for the management of EDS in PD patients.

The prevalence of RLS in Chinese patients with PD is approximately 8–35%. RLS can be difficult to distinguish from similar disorders in a clinical setting; therefore, the full profile of RLS must be investigated to establish a diagnosis. Importantly, there are numerous confounders that can result in false positives for RLS, such as dystonia, akathisia, painful neuropathy, and biphasic dyskinesia. Further studies are needed to understand the potential overlap between the symptoms, co-occurrence, and temporal order of occurrence of RLS in patients with PD, as well as to investigate the conversion of RLS to PD [8].

The SDB is grouped into obstructive sleep apnea (OSA) disorders, central sleep apnea disorders, sleep related hypoventilation disorders, and sleep-related hypoxemia disorder. The prevalence of OSA in patients with PD is estimated to be around 20–60%. It is characterized by upper airway narrowing or closure during sleep while respiratory effort continues. The consequences of OSA include cardiac arrhythmias, nighttime confusion, excessive daytime sleepiness, and functional decline. Some studies have found that OSA worsens cognitive functioning. Although the most effective treatment for OSA is continuous positive airway pressure, this may not result in overall cognitive improvement in patients with PD [9]. Hence, future research should include development of screening tools and better management of this disorder.

The 2017 Nobel Prize in Physiology or Medicine was awarded to elucidate of the molecular mechanisms controlling circadian rhythms. Circadian rhythms are biological rhythms that can affect mood, cognition, and autonomic and motor functions. In addition, there is increasing evidence of circadian disruption in PD. The circadian rhythm disorder is caused by alterations of the circadian timekeeping system; hence, understanding how circadian rhythms function, and which mechanisms can affect them, offers an opportunity to explore the pathophysiology of PD and potential treatments. For example, a recent *JAMA Neurology* article introduced light therapy as a treatment for the disturbed sleep and wakefulness associated with PD [10]. Light therapy caused an alteration in circadian rhythms wherein the light stimulated melanopsin-containing retinal ganglion cells via the retinohypothalamic tracts. This suggests that, while chronobiology has long been neglected, in the future we need to build systematic clinical investigations of circadian rhythm disruption in PD.

In general, management of sleep disorders in patients with PD is complex as these conditions are heterogeneous; therefore, treatment plans must be individualized and directed at the underlying cause(s) [11]. Prior to treatment, a comprehensive battery of clinical, neuropsychological, neuroimaging, and electro-physiological assessments should be conducted, and the sleep disorder cause and subtype need to be carefully evaluated. As examples, patients with PD that also experience insomnia should be treated based on the defined etiology (e.g., akinesia and drugs), whereas EDS often occurs secondarily as a symptom of another sleep disorder and can be treated with drugs, surgery, and/or increased nocturnal sleep. Additionally, when PD occurs in conjunction with RLS, other secondary factors and contributing comorbidities should be excluded, such as metabolic disorders (e.g., iron, folic acid, and vitamin B12 deficiency), end-stage renal disease, diabetes, pregnancy, and serotonergic antidepressants, and for RBD, the institution of appro-priate safety measures is a key component of any management plan.

Importantly, the influence of dopaminergic and other PD medications on sleep need to be accounted for when designing a treatment plan. Each of the sleep disor-ders discussed above can potentially be affected (either positively or negatively) by antiparkinsonian medications. For example, dopaminergic medications, particularly dopamine agonists, affect subjective sleepiness, and many dopaminergic agents can be effective for treating RLS in patients with PD. In addition, dopaminergic therapy can improve dream-enactment behavior in PD patients with RBD.

As previously mentioned, circadian-based therapies, such as timed-light expo-sure and melatonin, should be a focus of current research. Additionally, promoting behavioral interventions including proper sleep hygiene habits, increased activity during the day, and restriction of daytime napping that improve the consolidation of sleep–wake cycles can also be effective [2]. To this end, management of nighttime sleep quality may also be beneficial for motor symptoms in patients with PD [11]. Finally, transcranial magnetic stimulation and deep brain stimulation are novel treatments that might improve sleep disorders.

While recent reports have shown a clear association between sleep and PD, it is still unclear if primary sleep disorders increase the risk of developing PD and/or enhance the rate of progression, or if they arise as a consequence of PD. In addition, there are numerous unanswered questions regarding effective diagnostic assess-ments and management of sleep disorders in these cases, as well as the epidemiol-ogy, pathophysiology, clinical impact, and implications of the underlying disease and its manifestations. Future work should address these issues using longitudinal studies that employ large cohorts of patients with PD and identify high-risk patients for neuroprotective interventions.

The field of research focusing on sleep and PD has made enormous and exciting strides, and it is likely that future research at a molecular level will provide better therapies in this progressive disease. Therefore, the genetic study of sleep-related disorders still lags behind other medical fields. At present, the challenges of this field are improving our understanding of sleep–wake regulation and function, dis-seminating knowledge on sleep and sleep disorders to physicians and the general population, and educating neurologists with the aim of improved diagnosis of sleep disorders for early detection. Finally, we need to develop better treatment options and devise up-to-date guidelines for the management of PD patients with primary or comorbid sleep disorders.

References

1. Falup-Pecurariu C, Diaconu S. Sleep dysfunction in Parkinson's disease. Int Rev Neurobiol. 2017;133:719–42. https://doi.org/10.1016/bs.irn.2017.05.033.
2. Videnovic A. Management of sleep disorders in Parkinson's disease and multiple system atrophy. Mov Disord. 2017;32(5):659–68. https://doi.org/10.1002/mds.26918.
3. Gan-Or Z, Alcalay RN, Rouleau GA, Postuma RB. Sleep disorders and Parkinson disease; lessons from genetics. Sleep Med Rev. 2018;41:101. https://doi.org/10.1016/j.smrv.2018.01.006.
4. Rios Romenets S, Creti L, Fichten C, Bailes S, Libman E, Pelletier A, et al. Doxepin and cognitive behavioural therapy for insomnia in patients with Parkinson's disease—a randomized study. Parkinsonism Relat Disord. 2013;19(7):670–5. https://doi.org/10.1016/j.parkreldis.2013.03.003.
5. Shen Y, Dai YP, Wang Y, Li J, Xiong KP, Mao CJ, et al. Two polysomnographic features of REM sleep behavior disorder: clinical variations insight for Parkinson's disease. Parkinsonism Relat Disord. 2017;44:66–72. https://doi.org/10.1016/j.parkreldis.2017.09.003.
6. Barber TR, Lawton M, Rolinski M, Evetts S, Baig F, Ruffmann C, et al. Prodromal parkinsonism and neurodegenerative risk stratification in REM sleep behavior disorder. Sleep. 2017;40(8):zsx071. https://doi.org/10.1093/sleep/zsx071.
7. St Louis EK, Boeve BF. REM sleep behavior disorder: diagnosis, clinical implications, and future directions. Mayo Clin Proc. 2017;92(11):1723–36. https://doi.org/10.1016/j.mayocp.2017.09.007.
8. Hogl B, Stefani A. Restless legs syndrome and periodic leg movements in patients with movement disorders: specific considerations. Mov Disord. 2017;32(5):669–81. https://doi.org/10.1002/mds.26929.
9. Harmell AL, Neikrug AB, Palmer BW, Avanzino JA, Liu L, Maglione JE, et al. Obstructive sleep apnea and cognition in Parkinson's disease. Sleep Med. 2016;21:28–34. https://doi.org/10.1016/j.sleep.2016.01.001.
10. Videnovic A, Klerman EB, Wang W, Marconi A, Kuhta T, Zee PC. Timed light therapy for sleep and daytime sleepiness associated with Parkinson disease: a randomized clinical trial. JAMA Neurol. 2017;74(4):411–8. https://doi.org/10.1001/jamaneurol.2016.5192.
11. Chahine LM, Amara AW, Videnovic A. A systematic review of the literature on disorders of sleep and wakefulness in Parkinson's disease from 2005 to 2015. Sleep Med Rev. 2017;35:33–50. https://doi.org/10.1016/j.smrv.2016.08.001.

Zeitfracht Medien GmbH
Ferdinand-Jühlke-Straße 7
99095 Erfurt, Deutschland
produktsicherheit@kolibri360.de